I0462706

Structural Motion Control in MSC.NASTRAN

AUGUST 2010

SREEJIT RAGHU
MEng DIC ACGI MIStructE CEng MIEM

TABLE OF CONTENTS

ACKNOWLEDGEMENTS

My humble gratitude to the Almighty, to Whom this and all work is dedicated.

A special thank you also to my teachers at Imperial College of Science, Technology and Medicine, London and my fellow engineering colleagues at Ove Arup and Partners London and Ramboll Whitbybird London.

Sreejit Raghu

LIST OF SYMBOLS AND NOTATIONS

Elemental Notations

$\{y\}$	=	displacement function
$[N]$	=	shape functions matrix
$\{f\}$	=	element forces in element axes including fixed end forces
$\{d\}$	=	element deformation in element axes
$\{b\}$	=	element body forces
$[k]$	=	element constitutive matrix
$[m]$	=	element mass matrix
$[c]$	=	element viscous damping matrix
$\{p\}$	=	element nodal loading vector
$[T]$	=	transformation matrix
W	=	work done by external loads

SDOF, MDOF and Modal Dynamic Equation of Motion Notations

m	=	SDOF mass
$[M]$	=	Global MDOF mass matrix
c	=	SDOF viscous damping constant
$[C]$	=	Global MDOF viscous damping matrix
k	=	SDOF stiffness
$[K]$	=	Global MDOF stiffness matrix
u	=	SDOF displacement
$\{u\},\{U\}=$		Global MDOF displacement matrix
$\{P\}$	=	Global nodal loading vector
$M_i, [M]=$		Modal (generalized) mass and modal (generalized) mass matrix
$C_i, [C]=$		Modal (generalized) damping and modal (generalized) damping matrix
$K_i, [K]=$		Modal (generalized) stiffness and modal (generalized) stiffness matrix
$\xi_i, \{\xi_i\}=$		Modal displacement response and modal displacement response vector

SDOF Dynamic Notations

ω_n	=	Natural circular frequency, $(k/m)^{1/2}$
ω_d	=	Damped natural circular frequency, $\omega_n(1-\zeta^2)^{1/2}$
ω	=	Frequency of forcing function
c	=	Viscous damping constant
c_{cr}	=	Critical viscous damping constant, $2(km)^{1/2} = 2m\omega_n$
ζ	=	Damping ratio (fraction of critical), c/c_{cr}
δ	=	Logarithmic decrement

SDOF Free Vibrational Notations

G	=	Complex starting transient response function, $G = G_R + iG_I$

SDOF Time Domain Loading and Transient and Steady-State Response Notations

$P(t)$	=	Loading function
p_0	=	Force excitation amplitude
p_0/k	=	Static displacement
$D(t)$	=	Dynamic amplification factor

D_{max}	=	Maximum dynamic amplification factor
$u(t)$	=	Displacement response, $D(t)(p_0/k)$
u_{max}	=	Maximum displacement response, $D_{max}(p_0/k)$

Modal Time Domain Loading and Transient and Steady-State Notations

$\{P(t)\}$	=	Loading function vector
$P_i(t)$	=	Modal loading function, $P_i(t) = \{\phi_i\}^T \{P(t)\}$
p_{0i}	=	Modal force excitation amplitude
p_{0i}/K_i	=	Modal static displacement
$D_i(t)$	=	Modal dynamic amplification factor
$D_{i\,max}$	=	Modal maximum dynamic amplification factor
$\xi_i(t)$	=	Modal displacement response, $\xi_i(t) = D_i(t)p_{0i}/K_i$
$\xi_{i\,max}$	=	Modal maximum displacement response, $\xi_{i\,max} = D_{i\,max}\,p_{0i}/K_i$
$\{u(t)\}$	=	Displacement response vector, $\{u(t)\} = [\Phi]\{\xi(t)\}$

SDOF Frequency Domain Loading and Steady-State Response Notations

$P(t)$	=	SDOF Time domain harmonic loading function, $P(t) = \text{Real}\,[\,P(\omega)e^{i\omega t}\,]$
$P(\omega)$	=	SDOF frequency domain complex harmonic loading function
p_0	=	SDOF harmonic loading amplitude
p_0/k	=	SDOF static displacement
$D(\omega)$	=	SDOF (magnitude of the) dynamic amplification factor
$D_{resonant}$	=	SDOF (magnitude of the) dynamic amplification factor at resonance when $\omega = \omega_n$
D_{max}	=	SDOF maximum (magnitude of the) dynamic amplification factor when $\omega = \omega_n(1-2\zeta^2)^{1/2}$
$F(\omega)$	=	SDOF complex displacement response function (FRF), $F(\omega) = D(\omega)(p_0/k)e^{-i\theta}$
$H(\omega)$	=	SDOF transfer function, $H(\omega) = D(\omega)(1/k)e^{-i\theta}$
$F_{resonant}$	=	SDOF complex displacement response function at resonance, $F_{resonant} = D_{resonant}(p_0/k)e^{-i\theta}$
F_{max}	=	SDOF complex maximum displacement response function, $F_{max} = D_{max}(p_0/k)e^{-i\theta}$
$u(t)$	=	SDOF time domain displacement response, $u(t) = \text{Real}\,[\,F(\omega)e^{i\omega t}\,]$
$u_{resonant}$	=	SDOF time domain displacement response at resonance, $u(t) = \text{Real}\,[\,F_{resonant}\,e^{i\omega t}\,]$
u_{max}	=	SDOF time domain maximum displacement response, $u(t) = \text{Real}\,[\,F_{max}\,e^{i\omega t}\,]$
T_r	=	SDOF transmissibility of displacement, acceleration or force

Modal Frequency Domain Loading and Steady-State Response Notations

$\{P(t)\}$	=	Time domain harmonic loading function vector, $\{P(t)\} = \text{Real}\,[\,\{P(\omega)\}\,e^{i\omega t}\,]$
$\{P(\omega)\}$	=	Frequency domain complex harmonic loading function vector
$P_i(\omega)$	=	Modal frequency domain complex harmonic loading function vector, $P_i(\omega) = \{\phi_i\}^T \{P(\omega)\}$
p_{0i}	=	Modal harmonic loading amplitude
p_{0i}/K_i	=	Modal static displacement
$D_i(\omega)$	=	Modal (magnitude of the) dynamic amplification factor
$D_{i\,resonant}$	=	Modal (magnitude of the) dynamic amplification factor at resonance when $\omega = \omega_{ni}$
$D_{i\,max}$	=	Modal maximum (magnitude of the) dynamic amplification factor when $\omega = \omega_{ni}(1-2\zeta_i^2)^{1/2}$
$\xi_i(\omega)$	=	Modal complex displacement response function (FRF), $\xi_i(\omega) = D_i(\omega)p_{0i}/K_i\,e^{-i\theta i}$
$\xi_{i\,resonant}$	=	Modal complex displacement response function at resonance, $\xi_{i\,resonant} = D_{i\,resonant}\,p_{0i}/K_i\,e^{-i\theta i}$
$\xi_{i\,max}$	=	Modal complex maximum displacement response function, $\xi_{i\,max} = D_{i\,max}\,p_{0i}/K_i\,e^{-i\theta i}$
$\{u(t)\}$	=	Time domain displacement response vector, $\{u(t)\} = \text{Real}\,[\,[\Phi]\{\xi(\omega)\}e^{i\omega t}\,]$

Additional Abbreviations

ML: Materially Linear
MNL: Materially Nonlinear
GL: Geometrically Linear
GNL: Geometrically Nonlinear
[] = matrix
{} = column vector
<> = row vector

1.1   GL, ML Passive Structural Motion Control

Methods of passive structural motion control are: -

    i.   Optimum stiffness and mass distribution
    ii.   Isolation systems
    iii.   Optimum damping distribution
    iv.   Tuned mass damper systems

1.1.1   Optimum Stiffness and Mass Distribution

1.1.1.1   Concepts of Forced Frequency Response of Deterministic Periodic Harmonic Load Excitations

In a modal forced frequency response analysis, the modal response in modal coordinates ξ_i and the modal response in physical coordinates $u_i(t)$ are

$$\xi_i(\omega) = D_i(\omega)\frac{p_{0i}}{K_i}e^{-i\theta_i} = D_i(\omega)\frac{p_{0i}}{\omega_{ni}^2 M_i}e^{-i\theta_i} , \theta_i = \tan^{-1}\frac{2\zeta_i\omega/\omega_{ni}}{\left(1-\omega^2/\omega_{ni}{}^2\right)}$$

$$\{u_i(t)\} = \text{Re}\,al\left[\{\phi_i\}\xi_i(\omega)e^{i\omega t}\right]$$

The relative importance of each mode is encapsulated in the values of the **modal responses ξ_i. However, the physical response effects can be more critical from less significant modes (i.e. with lower modal responses) due to the inherent shape of the eigenvector even if the scaling factor (i.e. the modal response) is smaller.** The following parameters are of particular interest in controlling the modal response: -

    I.   The **modal mass** and **(the amplitude of the) modal force** must be considered together as their magnitudes are related to the arbitrary normalization of ϕ_i. MAX normalization scales the eigenvectors such that their maximum component is unity and all other components less than unity. The relative magnitude of the modal mass between different modes scaled by MAX is not in itself an indication of the relative importance of the particular mode as that also depends on the magnitude and location of the applied load excitation, given by the amplitude of the modal force. To reduce the modal response ξ_i, the modal mass should be maximized **AND** the amplitude of the modal force should be minimized. The modal mass will be maximized if there is **more structural mass** in the mode, hence the heavier the structure, the greater shall be the modal mass of most modes, and the lower the modal response. The amplitude of the modal force will be minimized if the **amplitude of the applied force is minimized** and if the **location of the applied force corresponds to the DOF with smaller components of the eigenvector**. This is obvious from a physical viewpoint, as clearly the modal response should be greater if the excitation is applied at the maximum locations of the eigenvector. With MAX normalization, the value of the modal mass is still of significance. A graph of (MAX normalized) modal mass versus modal frequency is most illustrative of the global and local modes. The greater the modal mass, the lower will be the response assuming constant amplitude of modal force. A very small modal mass (several orders of magnitude smaller than that of other modes) obtained from MAX normalization indicates a local mode or an isolated mechanism. If the applied force were to be located at the mechanism, the amplitude of the modal force may be significant, and hence so may the modal response. **If the distribution of mass is uniform, then higher modes of the same type (for instance higher bending modes relative to the fundamental bending mode) MAY have the same MAX normalized modal masses (simply supported beam modal masses 0.5mL and cantilever beam modal masses 0.25mL).** However other types of global modes of higher frequency than the fundamental frequency may have lower modal masses. For instance, the first torsional mode of a tower, which may be of a higher natural frequency than the first bending mode, may have a lower modal mass. This means that higher frequency global modes can also be significantly excited if the applied force was such that it corresponded to the greater values of their eigenvector.

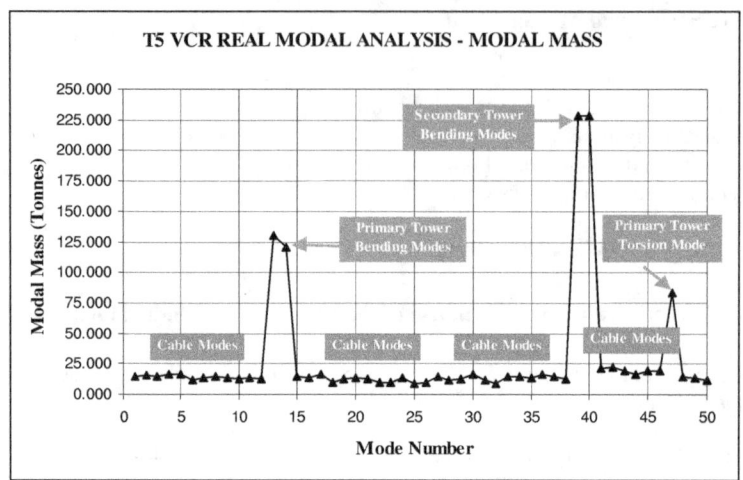

Hence, the modal masses of each and every mode are values that can be compared if all the modes had the same amplitude of modal force. The amplitude of the modal force for each and every mode can be made to be the same if there is only one concentrated applied force at one DOF and if the POINT normalization corresponding to that DOF is used for all modes. POINT normalization allows the user to choose a specific DOF at which the modal displacements are set to 1 or −1. Hence, the amplitude of the modal force for all modes will be the same and only the modal mass need to be compared. In this case, there will be significant difference in the order of magnitude of different global modes. The higher the modal mass with this method, the lower shall be the modal response. However, this normalization is not recommended because for complex structures, the chosen component may have very small values of displacement for higher modes causing larger numbers to be normalized by a small number, resulting in possible numerical roundoff errors and ridiculously higher modal masses. For instance, if the POINT normalization points to a DOF component which does not really exist in a particular mode, than all the other eigenvector terms will be normalized by a very small number, which will certainly result in numerical errors.

II.   For **ANY** natural mode, the (magnitude of the) dynamic amplification factor, $D_i(\omega)$ becomes infinite or is limited only by the modal damping of the particular mode when the excitation frequency approaches the natural frequency of the mode. The higher the modal damping, the lower the (magnitude of the) modal dynamic amplification factor, $D_i(\omega)$. Hence, the modal amplification can be reduced by either **mistuning the frequencies of excitation and the modal natural frequencies** or by **increasing the modal damping**. D_{imax} occurs when $\omega/\omega_n = (1-2\zeta^2)^{1/2}$.

$$D_i(\omega) = \frac{1}{\sqrt{\left(1 - \omega^2/\omega_{ni}^2\right)^2 + \left(2\zeta_i\omega/\omega_{ni}\right)^2}} \; ;$$

$$D_{i\,max} = \frac{1}{2\zeta\sqrt{\left(1 - \zeta^2\right)}}$$

III.   Higher modes are less significant then lower modes because of the **higher natural frequency** ω_{ni}^2 in the expression for the modal response ξ_i.

Hence, in order to reduce the total steady-state response (by controlling the distribution of mass, stiffness and damping) to a **fixed frequency and amplitude of excitation**, the following may be undertaken.

I.   The **structural mass distribution, stiffness distribution and boundary conditions** are modified so that there are no natural frequencies close to the frequency of excitation (and causing resonance) in order to minimize the (magnitude of the) dynamic amplification factor, $D_i(\omega)$. If the lowest natural frequency can be made significantly greater than the excitation frequency, then all the natural modes respond in a quasi-static manner to the excitation, and thus the response will only be dependent upon the amplitude of excitation as the frequency is low enough to be considered static. If this cannot be done, the natural modes should be tuned such that the excitation frequency is much greater than the lower natural frequencies. The response of the lower modes will be governed by inertial forces (the response of which is even lower than the quasi-static response) and the (magnitude of the) dynamic amplification factor, $D_i(\omega)$ will be minimized. However, $D_i(\omega)$ of higher modes will be more significant. But since ω_{ni}^2 features in the denominator of the modal response expression ξ_i, the higher modes will produce a lower modal response (although the modal physical response could be significant) when resonated.

II.   The **mass distribution** of the structure is increased in order to increase the modal masses (specifically of the modes with significant modal response ξ_i, but generally all), which in turn reduces the modal response ξ_i. This however will lower the natural frequencies and will affect the (magnitude of the) dynamic amplification factor, $D_i(\omega)$ of I.

III.   The **stiffness distribution** of the structure is increased in order to increase the natural frequencies ω_{ni}^2 (specifically of the modes with significant modal response ξ_i, but generally all), which feature in the denominator of the modal response ξ_i expression. This will however affect the (magnitude of the) dynamic amplification factor, $D_i(\omega)$ of I.

IV.   The **boundary conditions** of the structure are altered in order to increase the natural frequencies ω_{ni}^2 (specifically of the modes with significant modal response ξ_i, but generally all), which feature in the denominator of the modal response ξ_i expression. This will however affect the (magnitude of the) dynamic amplification factor, $D_i(\omega)$ of I.

V.   The **location and direction of the applied load excitation** is modified to not correspond to the large locations of the modes with significant modal response ξ_i in order to reduce the amplitude of the modal force, which in turn will reduce the modal response ξ_i.

VI.   The **damping distribution** is modified such as to maximize the modal damping of the modes with significant modal response ξ_i. This is done in order to minimize the (magnitude of the) dynamic amplification factor, $D_i(\omega)$ and hence minimizing the modal response ξ_i. If resonance is unavoidable, damping should be incorporated, as it is most efficient at resonance. For the design of explicit dampers, it is necessary to find the location and range of the damping parameter for optimum damping of a particular mode. The optimum location will clearly be at the position of the maximum components of the eigenvector. Optimum values of the viscous damping coefficient c (Ns/m) depend on the optimum modal damping obtained for a particular range of damping constant. There is always a plateau where the values of the damping constant c will give the optimum (highest) modal damping. This plateau range is obtained computationally by running repetitive complex modal analysis (MSC.NASTRAN SOL 107) with varying damping constants, and observing the range which gives the optimum damping for the structural mode that is to be damped.

1.1.1.2   Concepts of Forced Transient Response of Deterministic Load Excitations

In a modal forced transient response analysis, the modal response in modal coordinates ξ_i and the modal response in physical coordinates $u_i(t)$ are

$$\xi_i(t) = D_i(t)\frac{p_{0i}}{K_i} = D_i(t)\frac{p_{0i}}{\omega_{ni}^2 M_i}$$

$$\{u_i(t)\} = \{\phi_i\}\xi_i(t)$$

The relative importance of each mode is encapsulated in the values of the **modal responses ξ_i. However, the physical response effects can be more critical from less significant modes (i.e. with lower modal responses) due to the inherent shape of the eigenvector even if the scaling factor (i.e. the modal response) is smaller.** The following parameters are of particular interest in controlling the modal response: -

I.   The **modal mass** and **(the amplitude of the) modal force** must be considered together as their magnitudes are related to the arbitrary normalization of ϕ_i. MAX normalization scales the eigenvectors such that their maximum component is unity and all other components less than unity. The relative magnitude of the modal mass between different modes scaled by MAX is not in itself an indication of the relative importance of the particular mode as that also depends on the magnitude and location of the applied load excitation, given by the amplitude of the modal force. To reduce the modal response ξ_i, the modal mass should be maximized **AND** the amplitude of the modal force should be minimized. The modal mass will be maximized if there is **more structural mass** in the mode, hence the heavier the structure, the greater shall be the modal mass of most modes, and the lower the modal response. The amplitude of the modal force will be minimized if the **amplitude of the applied force is minimized** and if the **location of the applied force corresponds to the DOF with smaller components of the eigenvector**. This is obvious from a physical viewpoint, as clearly the modal response should be greater if the excitation is applied at the maximum locations of the eigenvector. With MAX normalization, the value of the modal mass is still of significance. A graph of (MAX normalized) modal mass versus modal frequency is most illustrative of the global and local modes. The greater the modal mass, the lower will be the response assuming constant amplitude of the modal force. A very small modal mass (several orders of magnitude smaller than that of other modes) obtained from MAX normalization indicates a local mode or an isolated mechanism. If the applied force were to be located at the mechanism, the amplitude of the modal force may be significant, and hence so may the modal response. **If the distribution of mass is uniform, then higher modes of the same type (for instance higher bending modes relative to the fundamental bending mode) MAY have the same MAX normalized modal masses (simply supported beam modal masses 0.5mL and cantilever beam modal masses 0.25mL).** However other types of global modes of higher frequency than the fundamental frequency may have lower modal masses. For instance, the first torsional mode of a tower, which may be of a higher natural frequency than the first bending mode, may have a lower modal mass. This means that higher frequency global modes can also be significantly excited if the applied force was such that it corresponded to the greater values of their eigenvector.

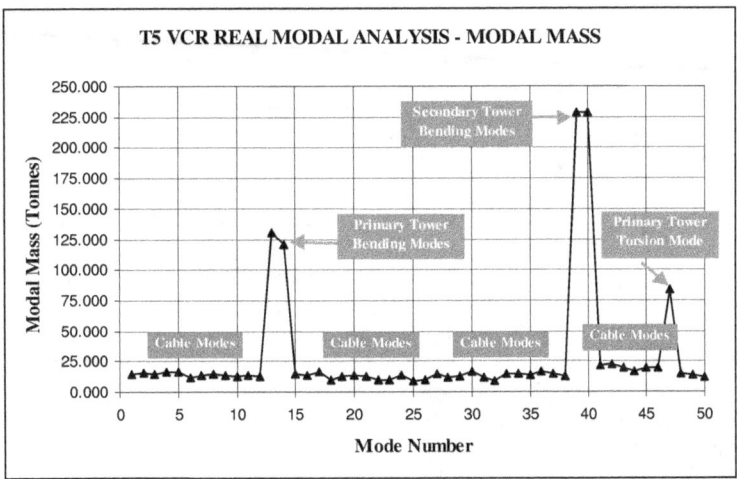

Hence, the modal masses of each and every mode are values that can be compared if all the modes had the same amplitude of modal force. The amplitude of the modal force for each and every mode can be made to be the same if there is only one concentrated applied force at one DOF and if the POINT normalization corresponding to that DOF is used for all modes. POINT normalization allows the user to choose a specific DOF at which the modal displacements are set to 1 or −1. Hence, the amplitude of the modal force for all modes will be the same and only the modal mass need to be compared. In this case, there will be significant difference in the order of magnitude of different global modes. The higher the modal mass with this method, the lower shall be the modal response. However, this normalization is not recommended because for complex structures, the chosen component may have very small values of displacement for higher modes causing larger numbers to be normalized by a small number, resulting in possible numerical roundoff errors and ridiculously higher modal masses. For instance, if the POINT normalization points to a DOF component which does not really exist in a particular mode, than all the other eigenvector terms will be normalized by a very small number, which will certainly result in numerical errors.

II. In general, $D_i(t)$ is a function of the natural circular frequency $\omega_{ni}^2 = K_i/M_i$ or the damped natural circular frequency ω_d, the time duration of loading t_d and the general time t. For a SDOF system, D_{imax} can be found by differentiation and is a function of the natural circular frequency $\omega_{ni}^2 = K_i/M_i$ or the damped natural circular frequency ω_d and the time duration of loading t_d. A graph of D_{imax} versus t_d / T_i is extremely illustrative of the maximum amplification that can be achieved for a particular type of impulsive loading.

Response Spectrum

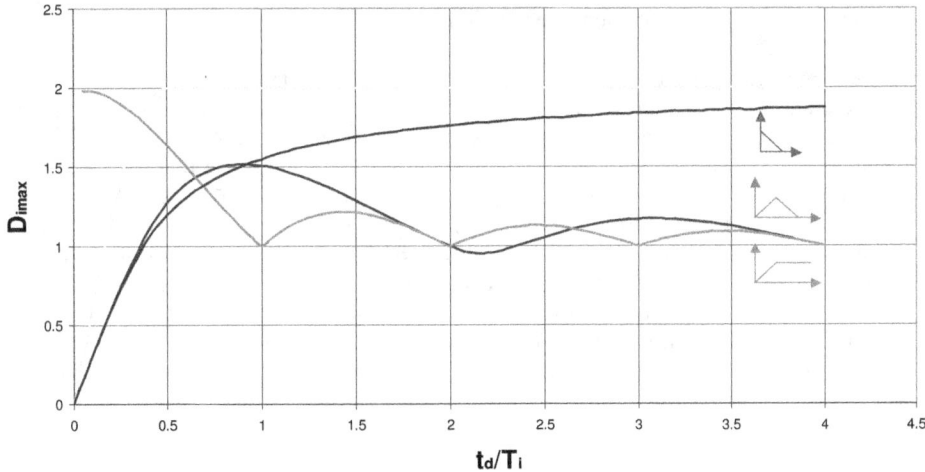

The above shows that the maximum dynamic amplification that can be produced from short duration impulse loadings is **two!** Maximum amplifications of higher modes are represented by points to the right of that for the fundamental mode.

On inspection of the red curve representing the **ramp up load** in duration t_d, it is apparent that when the ramp is of a shorter duration t_d than the fundamental period T_1 (i.e. $t_d/T_1 < 1.0$), the amplification is large decreasing from 2.0 to 1.0 (no dynamic amplification) when the duration equals the period. Higher modes are less amplified as t_d/T_i becomes larger. If the duration is larger than the fundamental period (i.e. $t_d/T_1 > 1.0$), the amplification will be small. The yellow curve with a large duration is simply a special case of the red curve with an instantaneous ramp (i.e. $t_d/T_1 = 0.0$), hence causing a dynamic amplification of 2.0. **In conclusion, if the ramp is such that it is gradual enough to be at least equal to the period of the fundamental mode (i.e. $t_d/T_1 > 1.0$), the amplification can be minimized.**

On inspection of the green curve representing the **impulsive loads**, if the duration of the impulse t_d is close to the period of the fundamental mode T_1, the amplification is large (~1.5). Higher modes are less amplified as t_d/T_i becomes larger. If the duration is much smaller than the fundamental period ($t_d/T_1 < \sim 0.4$), the amplification is less than unity, and hence not even the static response is observed as the impulse to too fast for the structure to react. If the duration is much larger than the fundamental period ($t_d/T_1 > 2.0$), the amplification is small. The yellow curve with a small duration is simply a special case of the green curve with an instantaneous impulse (i.e. $t_d/T_1 < \sim 0.2$), hence causing a dynamic amplification of less than 1.0 (i.e. less than static response). **In conclusion, if the impulsive load is much smaller in duration ($t_d/T_1 < \sim 0.4$) or much larger in duration ($t_d/T_1 > 2.0$) compared to the fundamental mode T_1, the amplification can be minimized.**

The fact that impulsive loads cause little amplification can also be quantified more directly than employing the dynamic amplification. The response from a true impulse ($t_1/T < \sim 0.2$) is readily obtained as the unit impulse function can be brought outside the Duhamel's or Convolution Integral. The maximum modal response (not the dynamic amplification $D_i(t)$) from an impulsive load ($t_1/T < 0.2$), can be obtained simply as follows.

$$\xi_i(\tau = t) = h(t - \tau) \int_{\tau=0}^{\tau=t} p(\tau)d\tau$$

$$\xi_{i\,max} = \frac{1}{M_i \omega_{ni}} \int_{\tau=0}^{\tau=t} \phi_i p(\tau)d\tau \qquad \text{undamped}$$

$$\xi_{i\,max} = \frac{1}{M_i \omega_{di}} \int_{\tau=0}^{\tau=t} \phi_i p(\tau)d\tau \qquad \text{damped}$$

where $\phi_i = $ modal component at excitation point

The above expression may seem to differ from the dynamic amplification approach where the denominator contains the modal stiffness, i.e. modal mass times the square of the natural circular frequency. This is because the above expression requires the explicit integration of the forcing function. The dynamic amplification approach differs in the sense that the integral (i.e. Duhamel's Integral) is already classically integrated to obtain the expression $D(t)$. Another equivalent interpretation of the above relationship is from the basic consideration of the conservation of momentum. The impacting particle (of small relative mass compared to mass of structure) imposes an impulse I onto the structure. The magnitude of I can be calculated as $m\Delta v$ where m is the small mass and Δv the change of velocity at impact. If there is no rebound Δv is the approach velocity. Conservation of momentum at impact requires the initial velocity of the structural mass to be I/M. A lightly damped system then displays damped free vibration with an initial displacement of approximately $I/(\omega_d M)$, or an initial velocity of approximately I/M or an initial acceleration of approximately $I\omega_d/M$.

The general derivation of the impulse force is a science in itself. Simple considerations can however result in good approximations. As mentioned, an impacting particle (of small relative mass compared to mass of structure) imposes an impulse $I = F\Delta t = m\Delta v$ onto the structure, where m and Δv are known. Knowing that $F\Delta t$ is the area under the impulse curve, making an estimate of the shape of the impulse curve and the duration Δt, we can thus estimate the peak amplitude. Hence, the impulse curve is defined.

Hence, if the structure is tuned such that the period of the fundamental mode is much less than the duration of the impulse ($t_1/T < 0.2$), the amplification will be very low and can readily be ascertained using the above relationship. Of course, higher modes may be amplified more than the fundamental mode but would respond less because of the larger natural frequency ω_{ni}^2.

III.   Higher modes are less significant then lower modes because of the **higher natural frequency ω_{ni}^2** in the expression for the modal response ξ_i.

Hence, in order to reduce the total (starting transient and steady-state) response (by controlling the distribution of mass, stiffness and damping) to a **ramped up loading or impulsive excitation**, the following may be undertaken.

I.   The **structural mass distribution, stiffness distribution and boundary conditions** are modified to minimize the dynamic amplification factor of the fundamental mode, $D_1(t)$. Higher modes may have a greater dynamic amplification. But since ω_{ni}^2 features in the denominator of the modal response expression ξ_i, the higher modes will produce a lower modal response (although the modal physical response could be significant).

II.   The **mass distribution** of the structure is increased in order to increase the modal masses (specifically of the modes with significant modal response ξ_i, but generally all), which in turn reduces the modal response ξ_i. This however will lower the natural frequencies and will affect the dynamic amplification factor, $D_i(t)$ of I.

III.   The **stiffness distribution** of the structure is increased in order to increase the natural frequencies ω_{ni}^2 (specifically of the modes with significant modal response ξ_i, but generally all), which feature in the denominator of the modal response ξ_i expression. This will however affect the dynamic amplification factor, $D_i(t)$ of I.

IV.   The **boundary conditions** of the structure are altered in order to increase the natural frequencies ω_{ni}^2 (specifically of the modes with significant modal response ξ_i, but generally all), which feature in the denominator of the modal response ξ_i expression. This will however affect the dynamic amplification factor, $D_i(t)$ of I.

V.   The **location and direction of the applied load excitation** is modified to not correspond to the large locations of the modes with significant modal response ξ_i in order to reduce the amplitude of the modal force, which in turn will reduce the modal response ξ_i.

1.1.1.3   Concepts of Forced Frequency Response of Deterministic Periodic Harmonic Base Excitations

For a SDOF system subjected to base harmonic excitations $u_0 \sin \omega t$, in absolute terms the response would be

$$u(t) = \frac{u_0 \sqrt{1 + (2\zeta\omega/\omega_n)^2}}{\sqrt{(1 - \omega^2/\omega_n^2)^2 + (2\zeta\omega/\omega_n)^2}} \sin(\omega t + \beta - \theta) \qquad \theta = \tan^{-1} \frac{2\zeta\omega/\omega_n}{(1 - \omega^2/\omega_n^2)}$$

Defining an expression for the relative transmissibility as the displacement response amplitude divided by the amplitude of the enforcing harmonic displacement u_0,

$$T_r = \frac{\sqrt{1 + (2\zeta\omega/\omega_n)^2}}{\sqrt{(1 - \omega^2/\omega_n^2)^2 + (2\zeta\omega/\omega_n)^2}}$$

This is the displacement transmissibility expression. The acceleration transmissibility is exactly similar. A plot of T_r versus ω/ω_n is somewhat similar to that of (magnitude of the) dynamic amplification D versus ω/ω_n, except that all the curves of different ζ pass through the same point of $T_r = 1.0$ when $\omega/\omega_n = \sqrt{2}$. Noting the curves after this point it is observed that damping tends to reduce the effectiveness of vibration isolation for frequency ratios greater than $\sqrt{2}$.

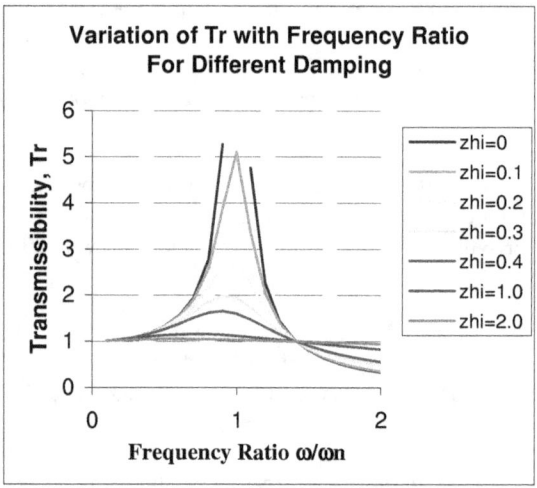

1.1.1.4   Concepts of Forced Transient Response of Deterministic Base Excitations

In a modal enforced motion transient response analysis, the modal response in modal coordinates ξ_i and the modal response in physical coordinates $u_i(t)$ are

$$\xi_i(t) = D_i(t)\frac{p_{0i}}{K_i} = -D_i(t)\frac{\{\phi_i\}^T[M]\{1\}\ddot{u}_0}{\omega_{ni}^2 M_i} = D_i(t)\Gamma_i\frac{\ddot{u}_0}{\omega_{ni}^2}$$

$$\text{where} \quad \Gamma_i = -\frac{\{\phi_i\}^T[M]\{1\}}{M_i} = -\frac{\{\phi_i\}^T[M]\{1\}}{\{\phi_i\}^T[M]\{\phi_i\}}$$

$$\{u_i(t)\} = \{\phi_i\}\xi_i(t)$$

Thus whereby we had the following for the modal response in load excitations

$$\xi_i(t) = D_i(t)\frac{p_{0i}}{K_i} = D_i(t)\frac{\{\phi_i\}^T\{p_0\}}{\omega_{ni}^2 M_i}$$

we now have the following for enforced base motion

$$\xi_i(t) = D_i(t)\frac{p_{0i}}{K_i} = -D_i(t)\frac{\{\phi_i\}^T[M]\{1\}\ddot{u}_0}{\omega_{ni}^2 M_i}$$

An **exceptionally crucial observation** is that whereby for load excitations the amplitude of the forcing vector $\{p_0\}$ may be sparse with only possibly one point with a value, that for enforced motion is quite different $[M]\{1\}\ddot{u}_0$ with all components having a value. Thus, the enforced motion amplitude of modal force is **special in the sense that the loading is uniformly distributed**. Hence, **higher modes will have lower amplitude of modal force because the positive and negative terms in the mode shape will cancel in the expression of the modal force**. Thus, higher modes will have a lower overall contribution (but inter-storey response may be significant). This does not generally occur for load excitations, unless the loading is uniformly distributed. To reiterate, higher modes in load excitations may well be significant if the location of the excitation is such that it corresponds to the greater components of the eigenvector and hence making the modal force is prominent. On the contrary, higher modes in base excitations will not be significant (to the total base shear but may well be significant for inter-storey response) because of the distributed nature of the forcing vector, canceling positive and negative terms in the eigenvector and making the modal force not prominent.

Because of the fact that the enforced based motion acceleration induces a force proportional to the mass and the base acceleration, clearly, higher masses would induce a greater force upon the structure.

In enforced motion, the **modal participation factor** is defined and is analogous to the ratio of the **load excitation amplitude of modal force p_{0i} but bereft of or normalized by the amplitude of the loading vector p_0** divided by the modal mass M_i.

1.1.1.5 Concepts of Forced Transient Response (Response Spectrum Analysis) of Random Non-Stationary Base Excitations

The **maximum** modal response in modal space for mode i is computed as follows

$$\xi_{i\,max}(t) = \Gamma_i S_D$$
$$\dot{\xi}_{i\,max}(t) = \Gamma_i S_V$$
$$\ddot{\xi}_{i\,max}(t) = \Gamma_i S_A$$

The **maximum** modal responses in physical space for mode i is then computed as follows

$$\{u_{imax}(t)\} = \{\phi_i\}\xi_{i\,max}(t) = \{\phi_i\}\Gamma_i S_D$$
$$\{\dot{u}_{imax}(t)\} = \{\phi_i\}\dot{\xi}_{i\,max}(t) = \{\phi_i\}\Gamma_i S_V$$
$$\{\ddot{u}_{imax}(t)\} = \{\phi_i\}\ddot{\xi}_{i\,max}(t) = \{\phi_i\}\Gamma_i S_A$$

An interesting point that could be made is the difference between the response spectrum method and the response due to enforced motion based on the dynamic amplification method. The **maximum** modal response from enforced base motion based on the dynamic amplification is

$$\xi_{i\,max}(t) = D_{i\,max}(t)\Gamma_i \frac{\ddot{u}_0}{\omega_{ni}^2}$$

Comparing, hence the spectral displacement is

$$S_D = D_{i\,max}(t)\frac{\ddot{u}_0}{\omega_{ni}^2}$$

1.1.2   Optimum Damping Distribution

Damping can often be ignored for short duration loading such as crash impulses or shock blast because the structure reaches peak before significant energy has had time to dissipate. Damping is important for long duration loadings such as earthquakes, wind and loadings such as rotating machines, which continually add energy to the structure and is especially critical when the response at resonance is to be established, for it is only the damping which balances the externally applied force as the potential energy (stiffness) and the kinetic energy (mass) cancel each other. Two mathematical formulations of damping exist, namely

(i)   **viscous damping c(du/dt)**, which is proportional to the response velocity (hence proportional to the forcing frequency), and

(ii)   **structural damping iGku**, which is proportional to the response displacement (hence independent of the forcing frequency)

The choice of the damping formulation should depend on the actual real world damping mechanism that is to be modeled. However, it is often the case that the formulation adopted is dependent upon that which can be handled by the dynamic solution algorithm. **Viscous** damping can be incorporated in both **time and frequency domain** solutions schemes whilst **structural** damping can only be incorporated in **frequency domain** solution schemes. The structural and viscous damping can be incorporated at 3 different levels, namely elemental, modal and global levels.

1.1.2.1   Elemental Damping Mathematical Models

Elemental damping damps all the structural modes of vibration accurately for all excitation frequencies irrespective of whether they are at resonance with the mode or not. Equivalent elemental damping introduces approximations.

1.1.2.1.1   Elemental Viscous Damping

This damping formulation can be implemented in the time domain (SOL 109, SOL 112, SOL 129) or in the frequency domain (SOL 108, SOL 111) with CDAMP, CVISC, CBUSH or CBUSH1D elements. It will damp all the structural modes of vibration accurately (obviously some modes will be damped more than others) for all excitation frequencies irrespective of whether they are at resonance with the mode or not.

The viscous damper is a device that opposes the relative velocity between its ends with a force that is proportional to velocity. The elemental viscous damping force is given by

$$f_d = c\,\dot{u}$$

The user specifies the elemental damping constant c in Ns/m, which is constant in a CDAMP, CVISC, CBUSH or CBUSH1D element.

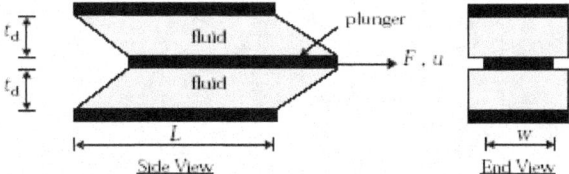

A material is considered to be elastic when the stresses due to an excitation are unique functions of the associated deformation i.e. $\tau = G_e \gamma$. Similarly, a material is said to be viscous when the stress state depends only on the deformation rates i.e. $\tau = G_v \dot{\gamma}$. Assuming no slip between the shear strain is related to the plunger motion by

$$\gamma = \frac{u}{t_d}$$

where t_d is the thickness of the viscous layer. Letting L and w represent the initial wetted length and width of the plunger respectively, the damping force is equal to

$$F = 2wL\tau$$

Substituting for τ

$$F = \left[\frac{2wL\,G_v}{t_d}\right]\dot{u}$$

Finally, for $f_d = c\,\dot{u}$

$$c = \left[\frac{2wL}{t_d}\right]G_v$$

The design parameters are the geometric measures w, L and t_d and the fluid viscosity, G_v.

For analysis, it is necessary to

(i)   **find the value of the damping parameters in practice to put into a computational model**

(ii)   **verify the damping parameters in the time or frequency domain computational model.**

These can be done by evaluating the response stress-strain curve of the element. The area under its response stress-strain curve represents the energy dissipated in an element.

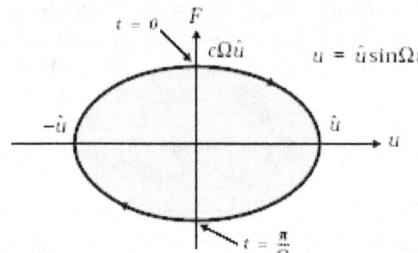

The displacement, velocity, acceleration and force are instantaneous quantities. Depicted for a SDOF system, if we know its velocity at any instant of time, the damping force at that instant is equal to the damping constant c multiplied by the velocity. Energy dissipated on the other hand is a cumulative quantity; hence the instantaneous force must be multiplied with an instantaneous differential displacement and summed over all the differential displacements. In the frequency domain, the energy dissipated by viscous damping during one cycle of harmonic vibration of forcing excitation $p_0 \sin\omega t$ is equal to the damping force multiplied the differential displacement, integrated over the single period of vibration. Hence,

$$\text{Displacement response amplitude, } D(\omega)p_0/k = \frac{p_0/k}{\sqrt{(1-\omega^2/\omega_n^2)^2 + (2\zeta\omega/\omega_n)^2}}$$

$$\text{Displacement response } u_p(t) = D(\omega)p_0/k\sin(\omega t - \theta)$$

$$\text{Velocity response } \dot{u}_p(t) = D(\omega)\omega p_0/k\cos(\omega t - \theta)$$

$$\text{Force response} = c\dot{u}_p(t)$$

$$\text{Energy dissipated in one harmonic cycle of } \omega,\ E_d = \int_{t=0}^{t=2\pi/\omega} c\dot{u}_p(t)du$$

$$= \int_{t=0}^{t=2\pi/\omega} c\dot{u}_p(t)\frac{du}{dt}dt$$

$$= \int_{t=0}^{t=2\pi/\omega} c\dot{u}_p^2(t)dt$$

$$= \int_{t=0}^{t=2\pi/\omega} c\big(D(\omega)\omega p_0/k\cos(\omega t - \theta)\big)^2 dt$$

$$= c\big(D(\omega)p_0/k\big)^2\omega^2 \int_{t=0}^{t=2\pi/\omega}\cos^2(\omega t - \theta)dt$$

$$= c\big(D(\omega)p_0/k\big)^2\omega^2\frac{\pi}{\omega}$$

$$= \pi\omega c\big(D(\omega)p_0/k\big)^2$$

The energy dissipated per cycle is directly proportional to the damping coefficient c, square of the response amplitude û and proportional to the driving frequency ω.

$$E_d = \pi \omega c \, \hat{u}^2$$

And the viscous damping coefficient is

$$\zeta = \frac{1}{4\pi} \frac{\Delta W}{W} = \frac{1}{4\pi} \frac{E_d}{W}$$

Shock absorbers in vehicle suspensions are viscous dampers. Fluid is forced through orifices located in the piston head as the piston rod position is changed, creating a resisting force which depends on the velocity of the rod. The damping coefficient can be varied by adjusting the control valve.

Radiation damping is a viscous damping mechanism occurring due to pressure wave radiation into medium surrounding the structure. This is a very efficient form of energy dissipation mechanism. If a plane wave is considered, then the energy transferred from the structure to the fluid per unit area per cycle is

$$E_d = \pi \omega \rho c (D(\omega)p_0/k)^2$$

Equating the energy dissipated per harmonic cycle for viscous damping and radiation damping,

$$E_{d \text{ viscous}} = \pi \omega c (D(\omega)p_0/k)^2 = E_{d \text{ radiation}} = \pi \omega \rho v (D(\omega)p_0/k)^2$$

$$c = \rho v$$

For a radiation plane wave, the viscous damping c per unit surface area of the structure is ρv where ρ is the density and v the speed of sound in the medium into which the structure propagates the energy. Radiation into solid medium thus gives a high value for this damping and hence accounts for most of foundation damping in buildings. It is also an efficient mechanism for structures surrounded in liquid.

For design, it is necessary to
      (i)   **find the range of the damping parameter for optimum damping of a particular mode**
For the design of explicit dampers, it is necessary to find the location and range of the damping parameter for optimum damping of a particular mode. The optimum location will clearly be at the position of the maximum components of the eigenvector. Optimum values of c depend on the optimum modal damping obtained for a particular range of damping constant. There is always a plateau where the values of the damping constant c will give the optimum (highest) modal damping. This plateau range is obtained computationally by running repetitive complex modal analysis (MSC.NASTRAN SOL 107) with varying damping constants, and observing the range which gives the optimum damping for the structural mode that is to be damped.

1.1.2.1.2   Elemental Structural Damping

This damping formulation can be implemented ONLY in the frequency domain (SOL 108, SOL 111) on the MAT1 card or using CELAS or CBUSH elements. It will damp all the structural modes of vibration accurately (obviously some modes will be damped more than others) for all excitation frequencies irrespective of whether they are at resonance with the mode or not.

The structural damping force is proportional to displacement and is in phase with velocity. The structural damping force is given by

$$f_d = iG_E k u$$

In the frequency domain, the user specifies the damping loss factor G_E and k in a MAT1, CELAS or CBUSH element.

For analysis, it is necessary to
      (i)   **find the value of the damping parameters in practice to put into a computational model**
      (ii)   **verify the damping parameters in the time or frequency domain computational model.**
These can be done by evaluating the response stress-strain curve of the element. The area under its response stress-strain curve represents the energy dissipated in an element.

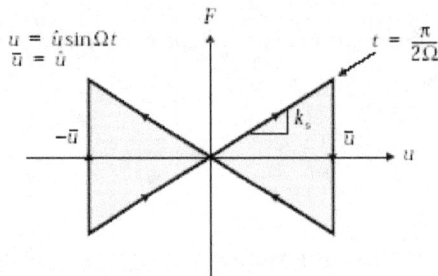

It can be shown that the energy dissipated in one harmonic cycle of ω by a structural damping element is

$$E_d = \pi G_E k \, \hat{u}^2$$

Structural damping is actually a **general friction damping mechanism** which allows for a variable magnitude of the friction force since it is proportional to the response displacement. This is contrasted with the less general friction damping mechanism i.e. Coulomb damping where the magnitude of the friction force is constant and limited.

1.1.2.1.3 Elemental Equivalent Viscous Damping for Structural Damping

There no reason to implement this damping formulation in the frequency domain as elemental structural damping can be specified instead. It can be implemented in the time domain (SOL 109, SOL 112, SOL 129) on the MAT1 card or using CELAS or CBUSH elements, all with the specification of the 'tie in' frequency PARAM, W4. If PARAM, W4 is not specified, no equivalent viscous damping for structural damping will be implemented. This damping formulation will NOT damp any of the structural modes of vibration accurately except the solitary 'tied in' mode that will be damped accurately ONLY when the excitation frequency matches its natural frequency (i.e. at resonance). Hence, this damping formulation should NOT be used, a modal approach should be used for the time domain instead.

Structural damping needs to be converted to equivalent viscous damping for solutions in the time domain by dividing the structural damping by the excitation frequency.

$$f_d = \frac{1}{\omega} G_E k \dot{u}$$

Although this equivalence can be implemented exactly in the frequency domain with the frequency dependent damping capability of the CBUSH element, this is unnecessary as structural damping can be defined in the frequency domain anyway. In the time domain, this requires the definition of a 'tie in' forcing frequency, chosen to correspond to a structural mode that is most prominent for the response (i.e. that which shows the greatest modal response with respect to time, a characteristic that can be obtained using SDISP, SVELO or SACCE in the modal solution), i.e. $\omega = \omega_1$. Hence, in the time domain, the user specifies the G_E, k and ω_1 from which $c = G_E k / \omega_1$ in a MAT1, CELAS or CBUSH element, all with PARAM, W4.

$$f_d = \frac{1}{\omega_1} G_E k \dot{u}$$

For analysis, it is necessary to
    (i)   **find the value of the damping parameters in practice to put into a computational model**
    (ii)   **verify the damping parameters in the time or frequency domain computational model.**
These can be done by evaluating the response stress-strain curve of the element. The area under its response stress-strain curve represents the energy dissipated in an element.

The equivalence is obtained by equating the energy dissipated per harmonic cycle for viscous damping and structural damping as follows.

$$E_{d \; viscous} = \pi \omega c (D(\omega) p_0 / k)^2 = E_{d \; structural} = \pi G_E k (D(\omega) p_0 / k)^2$$
$$c_{eq} = G_E k / \omega$$

1.1.2.1.4   Elemental Equivalent Viscous Damping for Velocity-Squared Damping

This damping formulation can be implemented approximately in the frequency domain (SOL 108, SOL 111) with CBUSH elements. It can be implemented approximately in the time domain (SOL 109, SOL 112) with CDAMP, CVISC, CBUSH or CBUSH1D elements. It should not be used in a nonlinear time domain solution (SOL 129) as the nonlinear (velocity squared) viscous damping can be modeled exactly in the nonlinear solution scheme with CBUSH1D elements.

Velocity-squared damping is directly proportional to the square of the velocity and opposes the direction of motion.

$$f_d = a\dot{u}^2$$

In a nonlinear time domain solution SOL 129, the user specifies 'a' in CBUSH1D. In the frequency domain SOL 108 and SOL 111, the user specifies 'a', variable ω and a constant displacement response amplitude $\hat{u}$ from which $c_{eq} = (8/3\pi)a\,\hat{u}\,\omega$ in an excitation frequency dependent CBUSH.

$$f_d = \frac{8}{3\pi} a\hat{u}\omega\dot{u}$$

In the linear time domain SOL 109 and SOL 112, the user specifies 'a', a constant displacement response amplitude $\hat{u}$ and a constant forcing frequency ω, the latter of which is chosen to correspond to a structural mode that is most prominent for the response (i.e. that which shows the greatest modal response with respect to time, a characteristic that can be obtained using SDISP, SVELO or SACCE in the modal solution), i.e. $\omega = \omega_1$ from $c_{eq} = (8/3\pi)a\,\hat{u}\,\omega_1$ in a CDAMP, CVISC, CBUSH or CBUSH1D element.

$$f_d = \frac{8}{3\pi} a\hat{u}\omega_1\dot{u}$$

For analysis, it is necessary to
      (i)   **find the value of the damping parameters in practice to put into a computational model**
      (ii)   **verify the damping parameters in the time or frequency domain computational model.**
These can be done by evaluating the response stress-strain curve of the element. The area under its response stress-strain curve represents the energy dissipated in an element.

The equivalence is obtained by equating the energy dissipated per harmonic cycle for viscous damping and velocity-squared damping as follows. For a linear viscous damper, the steady-state energy dissipated during one harmonic cycle is

$$E_d = \pi c\omega\,\hat{u}^2$$

where $\hat{u}$ is the amplitude of steady-state displacement, and ω is the excitation frequency i.e. the steady state response frequency. For a *linearized* velocity squared damper, steady-state energy dissipated during harmonic cycle

$$E_d = (8/3)a\omega^2\,\hat{u}^3$$

where a is the velocity-squared damping constant (Ns^2/m^2), and $\hat{u}$ is the amplitude of the steady-state displacement response. By equating the dissipated energies, the equivalent viscous damping coefficient for the velocity-squared damper is $c_{eq} = (8/3\pi)a\,\hat{u}\,\omega$.

Industrial hydraulic dampers are velocity-squared dampers. Velocity squared damping models the behavior observed when systems vibrate in fluid or when a fluid is rapidly forced through an orifice.

For design, it is necessary to
      (i)   **find the range of the damping parameter for optimum damping of a particular mode**
For the design of explicit dampers, it is necessary to find the location and range of the damping parameter for optimum damping of a particular mode. The optimum location will clearly be at the position of the maximum components of the eigenvector. Optimum damping constant 'a' can be obtained from repetitive SOL 129 analyses. Alternatively, assuming $\hat{u}$ and choosing $\omega = \omega_1$ so that $c_{eq} = (8/3\pi)a\,\hat{u}\,\omega_1$, repetitive SOL 107 can be run for numerous 'a' values and hence c_{eq}, until the optimum damping plateau is obtained.

1.1.2.1.5   Elemental Equivalent Viscous Damping for Coulomb Damping

This damping formulation can be implemented approximately in the frequency domain (SOL 108, SOL 111) with CBUSH elements. It can be implemented approximately in the time domain (SOL 109, SOL 112) with CDAMP, CVISC, CBUSH or CBUSH1D elements. It should not be used in a nonlinear time domain solution (SOL 129) as the damping can be modeled exactly in the nonlinear solution scheme with the CGAP (contact and) friction element.

It is assumed that the (constant and limited) force resisting the direction of motion is proportional to the normal force F_N between the sliding surfaces, independent of the magnitude of velocity but is in phase with the velocity (hence the sign of velocity term).

$$F = \overline{F} \operatorname{sgn}(\dot{u})$$

where $\overline{F}$ is the limiting Coulomb force magnitude and is equal to μF_N where F_N is the normal contact force and μ the coefficient of friction. In a nonlinear time domain solution SOL 129, the user specifies the coefficient of friction, μ in a CGAP contact-friction element. In the frequency domain SOL 108 and SOL 111, the user specifies μ, F_N, variable ω and variable $\hat{u}$ from which c is calculated from $c_{eq} = 4\mu F_N/(\pi\omega\hat{u})$. Due to the nature of the Coulomb force being in phase with the velocity, an equivalent viscous damping model can easily be used in a CBUSH element.

$$f_d = \frac{4\mu F_N}{\pi\omega\hat{u}}\dot{u}$$

where F_N is the normal force, μ the coefficient of friction and $\hat{u}$ the harmonic response amplitude. In the linear time domain SOL 109 and SOL 112, a constant displacement response $\hat{u}$ and forcing frequency ω is required, the latter of which is chosen to correspond to a structural mode that is most prominent for the response (i.e. that which shows the greatest modal response with respect to time, a characteristic that can be obtained using SDISP, SVELO or SACCE in the modal solution), i.e. $\omega = \omega_1$. In the time domain, the user specifies μ, F_N, ω_1 and $\hat{u}$ from which $c_{eq} = 4\mu F_N/(\pi\omega_1\hat{u})$ in a CDAMP, CVISC, CBUSH or CBUSH1D element.

$$f_d = \frac{4\mu F_N}{\pi\omega\hat{u}}\dot{u}$$

For analysis, it is necessary to
      (i)   **find the value of the damping parameters in practice to put into a computational model**
      (ii)   **verify the damping parameters in the time or frequency domain computational model.**
These can be done by evaluating the response stress-strain curve of the element. The area under its response stress-strain curve represents the energy dissipated in an element.

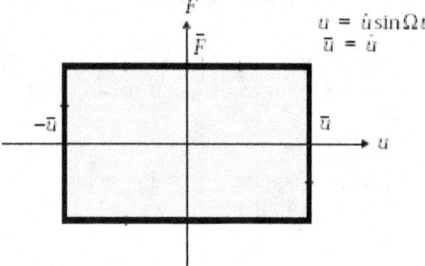

The energy dissipated is proportional to the friction force and the amplitude of vibration as follows.

$$E_d = 4\mu F_N \hat{u}$$

Coulomb damping or dry friction damping is found wherever **sliding friction** is found in machinery, laminated springs or bearings. Coulomb damping is also used to model **joint damping** mechanisms. Joint damping is found in highly assembled structures.

1.1.2.1.6   Elemental Equivalent Structural Damping for Hysteretic Damping

This damping formulation can be implemented ONLY in the frequency domain (SOL 108, SOL 111) on the MAT1 card or using CELAS or CBUSH elements. It will damp all the structural modes of vibration accurately (obviously some modes will be damped more than others) for all excitation frequencies irrespective of whether they are at resonance with the mode or not. It should not be used in a nonlinear time domain solution (SOL 129) as the damping can be modeled exactly in the nonlinear solution scheme with an elastic-plastic hysteretic material model MATS1 (TYPE PLASTIC).

The form of the damping force-deformation relationship depends on the stress-strain relationship for the material and the make-up of the device. For an elastic-perfectly plastic material, the limiting values are F_y, the yield force, and u_y, the displacement at which the material starts to yield; k_h is the elastic damper stiffness. The ratio of the maximum displacement to the yield displacement is referred to as the ductility ratio and is denoted by μ. In a nonlinear time domain solution SOL 129, the user specifies the initial stiffness k_h, the yield force F_y and the ductility ratio μ or the ultimate strain in a MATS1 (TYPE PLASTIC). In the frequency domain SOL 108 and SOL 111, the user specifies the equivalent structural damping loss factor G_E for the hysteretic damper as

$$G_E = 2(\mu - 1)$$

where μ is the ductility ratio in a MAT1, CELAS or CBUSH element.

For analysis, it is necessary to

    (i)   **find the value of the damping parameters in practice to put into a computational model**

    (ii)   **verify the damping parameters in the time or frequency domain computational model.**

These can be done by evaluating the response stress-strain curve of the element. The area under its response stress-strain curve represents the energy dissipated in an element.

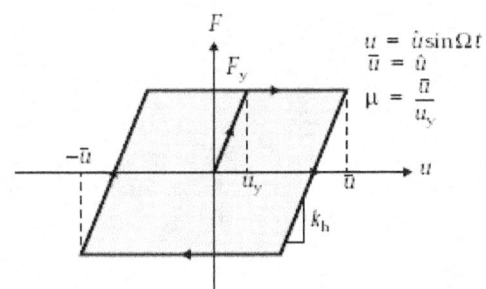

The work per cycle for hysteretic damping is

$$W_{hysteretic} = 4k_h u_y^2 [\mu - 1] = 4F_y \bar{u}\left[\frac{\mu - 1}{\mu}\right]$$

Hysteretic damping is due to the **inelastic deformation** of the material composing the device. Hysteretic damping is also offered by **hysteretic damper brace elements**.

For design, it is necessary to

    (i)   **find the range of the damping parameter for optimum damping of a particular mode**

For the design of explicit hysteretic damper brace elements, it is necessary to find the location and range of the damping parameter for optimum damping of a particular mode. The optimum location will clearly be at the position of the maximum components of the eigenvector. Optimum ductility ratio μ can be obtained from repetitive SOL 129 analyses. Alternatively, assuming repetitive SOL 107 can be run for numerous G_E values and hence μ, until the optimum damping plateau is obtained.

1.1.2.1.7   Elemental Equivalent Structural Damping for Viscoelastic Damping

This damping formulation can be implemented ONLY in the frequency domain (SOL 108, SOL 111) on the CBUSH elements or in SOL 108 with MAT1 and TABLEDi (with the SDAMPING Case Control Card). It should not be used in a nonlinear time domain solution (SOL 129) as the damping can be modeled exactly in the nonlinear solution scheme with the CREEP material model.

Viscoelastic materials are characterized by the shear modulus G (which relates to a stiffness k) and the loss factor, G_E both of which are temperature, strain and frequency dependent.

$$G(f) = G'(f) - iG''(f)$$

$$G' = \text{Shear storage modulus}$$

$$G'' = \text{Shear loss modulus}$$

The ratio

$$\frac{G''(f)}{G'(f)} = \tan\phi$$

is the shear loss tangent. A material is considered to be elastic when the stresses due to an excitation are unique functions of the associated deformation i.e. $\tau = G_e\gamma$. Similarly, a material is said to be viscous when the stress state depends only on the deformation rates i.e. $\tau = G_v\dot{\gamma}$.

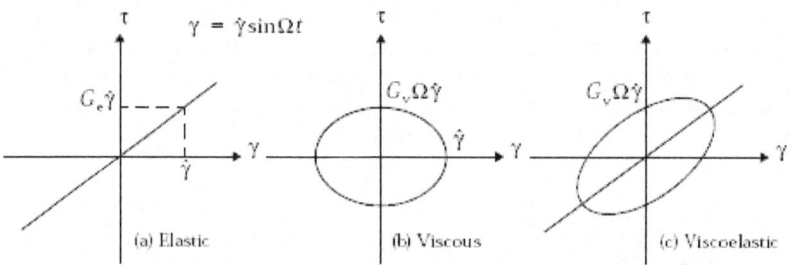

(a) Elastic   (b) Viscous   (c) Viscoelastic

An elastic material responds in phase with excitation whilst a viscous element responds 90 degrees out of phase. Hence a viscoelastic element responds 0 – 90 degrees out of phase with the excitation. In the frequency domain SOL 108 and SOL 111, the user specifies the excitation frequency dependent stiffness, $k(\omega)$ and the excitation frequency dependent loss factor, $G_E(\omega)$ in a CBUSH element. Alternatively in SOL 108,

SDAMPING = n to reference TABLEDi Bulk Data entry
MAT1 with G = G_{REF} (reference modulus) and GE = g_{REF} (reference element damping)
A TABLEDi Bulk Data entry with an ID = n is used to define the function TR(f)
A TABLEDi Bulk Data entry with an ID = n+1 is used to define the function TI(f)

where the stiffness matrix of the viscoelastic elements

$$[K_{dd}(f)]_V = \left[\frac{G'(f) + iG''(f)}{G_{REF}}\right][K_{dd}^1]_V$$

$$TR(f) = \frac{1}{g_{REF}}\left[\frac{G'(f)}{G_{REF}} - 1\right]$$

$$TI(f) = \frac{1}{g_{REF}}\left[\frac{G''(f)}{G_{REF}} - g\right]$$

For analysis, it is necessary to

(i)   find the value of the damping parameters in practice to put into a computational model

(ii)   verify the damping parameters in the time or frequency domain computational model.

These can be done by evaluating the response stress-strain curve of the element. The area under its response stress-strain curve represents the energy dissipated in an element.

Viscoelastic damping is present in most materials, albeit at very small levels. This is of course present in both the elastic and plastic range of the material deformation and their values are presented below.

Material	Loss Factor, G_E
Steel	0.001 – 0.008
Cast iron	0.003 – 0.03
Aluminium	0.00002 – 0.002
Lead	0.008 – 0.014
Rubber	0.1 – 0.3
Glass	0.0006 – 0.002
Concrete	0.01 – 0.06

Viscoelastic damping is utilized in **constrained layer damping** for structural and aeronautical applications. Examples include SUMITOMO-3M, SWEDAC and bituthene. This method works best when the constrained layer is put into shear and therefore the most effective location of such a layer is at the neutral axis of the floor beam or slab where the complementary shear stresses are high. Typical visco-elastic materials are fairly flexible and can exhibit significant creep. Therefore introduction of a constrained layer means that the structure is more flexible than it would be if it were monolithic. For static loads therefore these types of layers are not ideal, and so a compromise must often be reached. Constrained layer damping has quite often been applied to excessively lively existing structures by adding an extra "cover plate" on top of or beneath the existing floor in order to constrain a layer of visco-elastic material.

Slab

Truss end

The stiffness k of a (zero-length two orthogonal horizontal direction) spring representing an area A of viscoelastic constrained layer material is AG/t where t is the thickness and G the shear modulus. The variation of G and the loss factor G_E with temperature, excitation frequency has been established for bituthene.

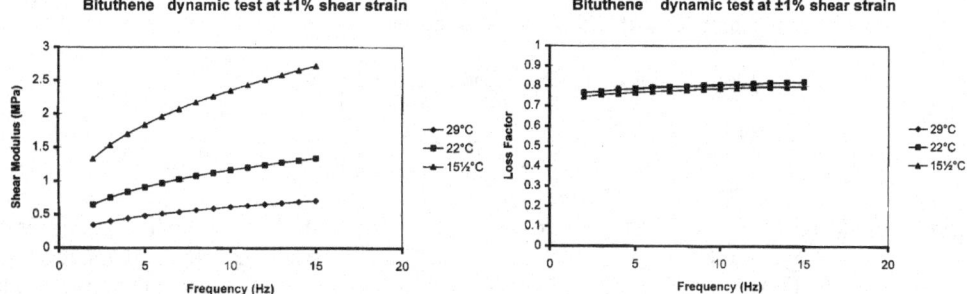

Temperature has dramatic effect on G but little on G_E. The material properties to use based on interpolation of test results at 20 °C and a frequency of 5Hz are thus

Shear Modulus   G   = 1.17 MPa
Loss factor G_E = 0.75 (at very small strains)

For design, it is necessary to
    (i)   **find the range of the damping parameter for optimum damping of a particular mode**
Optimum damping parameters k and G_E can be obtained from repetitive SOL 107 analyses, until the optimum damping plateau is obtained. For constrained layer viscoelastic damping, repetitive SOL 107 can be run until the value of k (and hence the thickness t since k = AG/t) that yields the optimum damping is obtained.

1.1.2.2 Modal Damping Mathematical Models

Modal damping damps all the structural modes of vibration specified with a modal damping accurately at all excitation frequencies irrespective of whether they are at resonance with the mode or not. Equivalent modal damping damps the structural modes of vibration specified with a modal damping accurately ONLY when the excitation frequency matches the frequency of each particular natural mode.

Note that the modal damping approach will not be valid if the modes are significantly complex, i.e. different parts of the structure are out-of-phase (reaching their maximum at different instances) in a particular mode as the modal damping values are applied on the real modes. Damping elements with high values of damping can result in the structure having totally different damped modes of vibration. For instance, a cable with a damper element with a very high damping coefficient attached to its center will behave as two separate cables, the damper element effectively producing a fixity at the center of the cable. This changes the fundamental mode shape of the cable from one of a low frequency to two cable mode shapes of higher fundamental frequency.

For analysis, it is necessary to
> (i)   **find the value of the damping parameters in practice to put into a computational model**
> (ii)   **obtain modal damping values from a set of explicit element dampers**

There are various methods of obtaining the modal damping coefficients in practice and from finite element models damped with individual explicit viscous or structural damping elements. **Complex modal analysis** (SOL 107) can be performed to ascertain the structural modal damping coefficients G_i for each and every mode. Alternatively, a **forced frequency response analysis** (SOL 108 or SOL 111) can be performed with a harmonic loading from which the modal damping coefficient can be ascertained in a variety of methods from the FRF. The methods of obtaining the damping from the FRF are: -

(i)   the half-power bandwidth method which approximates the modal damping in lightly-damped structures ($\zeta < 0.1$) where the ith modal damping coefficient is

$$\zeta_i = \frac{f_{i2} - f_{i1}}{2f_i}$$

where f_{i1} and f_{i2} are the beginning and ending frequencies of the half-power bandwidth defined by the maximum response divided by $\sqrt{2}$. The half-power bandwidth method assumes a SDOF response.

(ii)   the (magnitude of the) dynamic amplification factor at resonance method

$$\frac{1}{2\zeta_i} = \text{Dynamic Amplification Factor At Resonance of ith mode, } D_{i\,resonant}$$

$$= \frac{\text{Maximum Displacement Response Amplitude of ith mode, } D_{i\,resonant}\,p_0/k}{\text{Static Displacement Response Amplitude, } p_0/k}$$

Another method of obtaining modal damping estimates is the **logarithmic decrement method**. In order to obtain the values of ζ_i, a transient analysis (SOL 109, SOL 112, SOL 129) can be performed and successive peaks can be observed from the free vibration response. The free vibrational analysis can be set up using an initial displacement condition or an impulsive force that ramp up and down quickly with respect to the dominant period of the response and/or other natural frequencies of interest. The logarithmic decrement, δ is the natural log of the ratio of amplitude of two successive cycles of free vibration. Then the damping ratio is

$$\zeta \approx \frac{\delta}{2\pi}$$

The logarithmic decrement method assumes a SDOF response. Modal damping of higher modes requires signal filtering of the response time history.

1.1.2.2.1 Modal Viscous Damping

This damping formulation can be implemented in the modal time domain (SOL 112) with the SDAMPING Case Control Command and the TABDMP1 entry. It can also be implemented in the modal frequency domain (SOL 111) with SDAMPING Case Control Command, the PARAM, KDAMP and the TABDMP1 entry. It will damp all the structural modes of vibration specified with a modal damping value accurately for all excitation frequencies irrespective of whether they are at resonance with the mode or not.

The damping force is given by

$$f_{id} = \zeta_i \, (2M_i \omega_{n_i}) \, (i\omega \xi_i(\omega))$$

The user specifies ζ_i.

1.1.2.2.2 Modal Structural Damping

This damping formulation cannot be implemented in the modal time domain (SOL 112). It can be implemented in the modal frequency domain (SOL 111) with the SDAMPING Case Control Command, the PARAM, KDAMP and the TABDMP1 entry where it will damp all the structural modes of vibration specified with a modal damping value accurately for all excitation frequencies irrespective of whether they are at resonance with the mode or not.

The damping force is given by

$$f_{id} = iG_i K_i \xi_i(\omega)$$

The user specifies G_i.

1.1.2.2.3 Modal Equivalent Viscous Damping for Modal Structural Damping

There no reason to implement this damping formulation in the modal frequency domain as modal structural damping can be specified instead. It can be implemented in the modal time domain (SOL 112) with the SDAMPING Case Control Command and the TABDMP1 entry. This damping formulation will damp all modes specified with a modal damping value accurately ONLY when the excitation frequency matches the natural frequency of the particular mode (i.e. at resonance), often an acceptable approximation.

This modal damping formulation is extremely useful in modelling structural damping in time domain solutions. Time domain solutions, be they direct or modal cannot handle structural damping. Incorporating elemental structural damping within a time domain solution requires a 'tie in' frequency, hence damping the different modes inaccurately. The best method would be to use a modal approach, i.e. SOL 112 where modal damping can be specified and internally automatically converted into modal viscous damping. All the modes are then damped accurately when the excitation frequency matches the natural frequency of the particular mode. Again, the method works if the modes are only slightly complex, as the modal damping estimates are applied onto the real modes.

The following presents the appropriate conversion between modal viscous and modal structural damping. It will be shown that the two forms of damping are only theoretically equivalent when the forcing frequency ω equals the natural frequency of the *structural* mode of vibration ω_n. This is proven mathematically and pictorially below showing the variation of viscous and structural damping with forcing frequency.

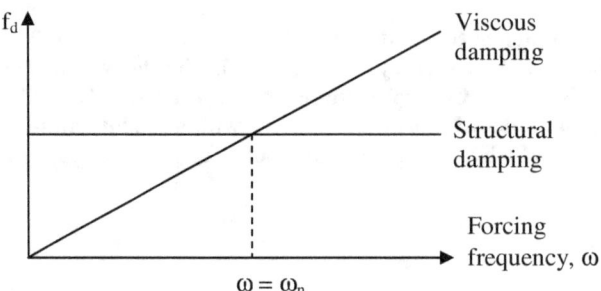

For the viscous damping force to equal the structural damping force

$$f_{\text{viscous}} = f_{\text{structural}}$$
$$c\dot{u} = iGku$$
$$ic\omega u = iGku$$

The damping constant c is specified as a percentage of critical damping ζ of a particular structural mode

$$i\omega(\zeta 2m\omega_n)u = iGku$$
$$\omega(\zeta 2m\omega_n) = Gk$$

If and only if $\omega = \omega_n$

$$\omega_n(\zeta 2m\omega_n) = Gk$$
$$\zeta 2m\omega_n^2 = Gk$$
$$\zeta 2mk / m = Gk$$
$$\zeta = G / 2$$

Although the two damping formulations are only identical the resonances $\omega = \omega_n$ and not for other forcing frequencies, damping is most critical at resonance when $\omega \approx \omega_n$ (when the inertial and stiffness forces cancel) and so the switch between viscous and structural formulations is often acceptable in practice. Whenever possible, the damping formulation chosen (i.e. whether viscous or structural) should be such as to simulate the actual real-world damping mechanism. Nevertheless, the use of viscous damping to simulate structural damping (and vice versa for that matter) is possible, but it must be borne in mind that the conversion is only accurate at resonance when the *structural* modal frequencies equal the forcing frequencies. From above, it is obvious that the use of a **viscous damping** formulation to represent a structural damping mechanism will **overestimate** damping when the forcing frequency ω is higher than ω_{ni} and **underestimate** damping when the forcing frequency ω is lower than ω_{ni}. Conversely, the use of a **structural damping** formulation to represent a viscous damping mechanism will **underestimate** damping when the forcing frequency ω is higher than ω_{ni} and **overestimate** damping when the forcing frequency ω is lower than ω_{ni}.

1.1.2.3   Global Damping Mathematical Models

Global damping formulations are the crudest method of modelling damping. This method does not damp all the natural modes accurately because it is often based on the damping estimates of either one or two modes only. These modal estimates are obtained using methods of obtaining modal damping values described previously.

1.1.2.3.1   Global Viscous Damping (Mass Proportional Damping)

The global (mass proportional) viscous damping force is given by

$$[F]_d = [C]\{\dot{u}\}$$
$$= \alpha[M]\{\dot{u}\}$$
$$= \zeta_i 2\omega_{ni}[M]\{\dot{u}\}$$

The user specifies α.

The values of ζ_i and ω_{ni} are based one mode only, usually the first fundamental mode which usually dominates the response. Hence, the method tends to underestimate damping in modes of higher than ω_{ni} frequencies and overestimate damping in modes of lower than ω_{ni} frequencies.

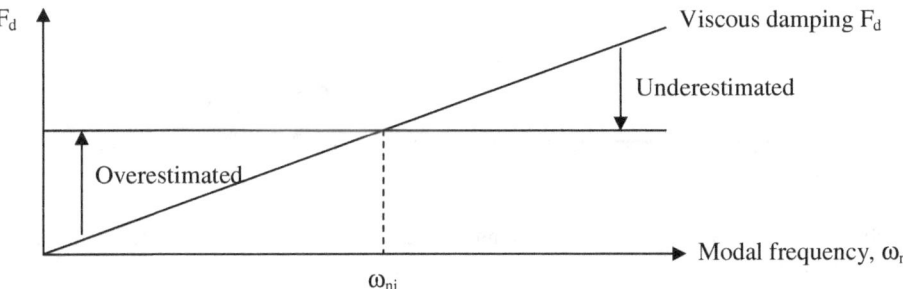

1.1.2.3.2   Global Structural Damping

This damping formulation can ONLY be implemented in the frequency domain (SOL 108, SOL 111) using PARAM, G. It will damp all the structural modes of vibration accurately (obviously some modes will be damped more than others) for all excitation frequencies irrespective of whether they are at resonance with the mode or not if and only if all the deformable elements have associated structural damping of the same value.

The global structural damping force is given by

$$[F]_d = iG[K]\{u\}$$

The user specifies G.

1.1.2.3.3   Global Equivalent Viscous Damping for Structural Damping (Stiffness Proportional Damping)

There no reason to implement this damping formulation in the frequency domain as elemental structural damping and/or modal structural damping can be specified instead. It can be implemented in the time domain (SOL 109, SOL 112, SOL 129) using PARAM, G with the specification of the 'tie in' frequency PARAM, W3. If PARAM, W3 is not specified, no equivalent viscous damping for structural damping will be implemented. This damping formulation will NOT damp any of the structural modes of vibration accurately

except the solitary 'tied in' mode that will be damped accurately ONLY when the excitation frequency matches its natural frequency (i.e. at resonance). Hence, this damping formulation should NOT be used, a modal approach should be used for the time domain instead.

The global (stiffness proportional) viscous damping force is given by

$$[F]_d = \beta[K]\{\dot{u}\}$$
$$= \frac{2\zeta_i}{\omega_{ni}}[K]\{\dot{u}\}$$

The user specifies β.

The values of ζ_i and ω_{ni} are based one mode only, usually the first fundamental mode which usually dominates the response. Hence, the method tends to overestimate damping in modes of higher than ω_{ni} frequencies and underestimate damping in modes of lower than ω_{ni} frequencies.

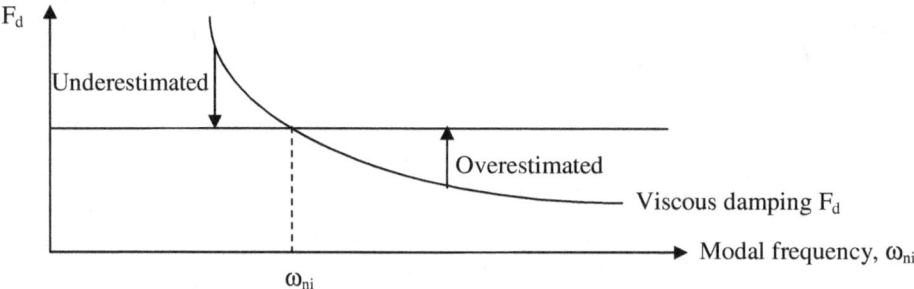

This method is crudely used to model the structural damping mechanism in a global viscous formulation. This is so because, from the above equation

$$[F]_d = \beta[K]\{\dot{u}\}$$
$$= \frac{2\zeta_i}{\omega_{ni}}[K]\{u\}i\omega$$

When $\omega = \omega_{ni}$

$$[F]_d = i2\zeta_i[K]\{u\}$$
$$= iG_i[K]\{u\}$$

As mentioned, a viscous formulation to model structural damping will overestimate damping when the forcing frequency ω is higher than ω_{ni} and underestimate damping when the forcing frequency ω is lower than ω_{ni}. This is even more exaggerated here as the damping estimate is based upon just one mode.

1.1.2.3.4 Global Equivalent Viscous Damping for Viscous and Structural Damping (Mass and Stiffness Proportional Rayleigh Damping)

The Rayleigh proportional viscous damping force is given by

$$[F]_d = [\alpha[M] + \beta[K]]\{\dot{u}\}$$

The user specifies α and β.

The factors α and β are chosen to give the correct damping at two excitation frequencies. For a specified damping factor ζ at an excitation frequency ω then

$$\zeta = \frac{1}{2}\left(\frac{\alpha}{\omega} + \beta\omega\right)$$

And using this at 2 frequencies ω_r and ω_s with the required damping factors ζ_r and ζ_s respectively gives

$$\alpha = 2\omega_r\omega_s \frac{\zeta_s\omega_r - \zeta_r\omega_s}{\omega_r^2 - \omega_s^2}$$

$$\beta = 2\frac{\zeta_r\omega_r - \zeta_s\omega_s}{\omega_r^2 - \omega_s^2}$$

The method is based on the damping coefficients ζ_r and ζ_s of two distinct modes ω_r and ω_s. By specifying $\zeta_r = \zeta_s$, all structural natural modes between modal frequencies ω_r and ω_s will have damping less than at modes ω_r and ω_s.

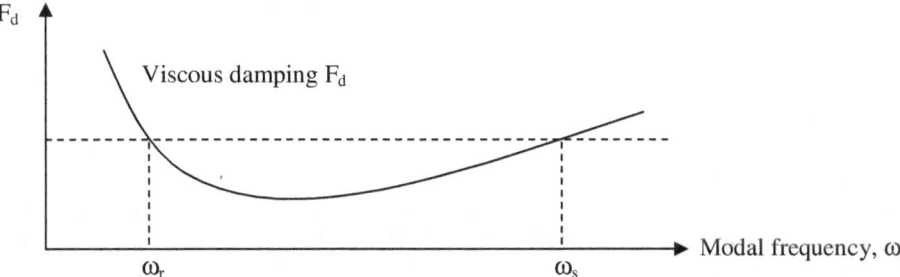

1.1.2.4 Damping Formulation Conclusions

The choice of using a viscous damping formulation or a structural damping formulation should depend on the actual real world damping mechanism being modeled. Elemental damping formulations damp all modes accurately at all excitation frequencies. If an equivalent elemental damping formulation is used, then not all natural modes will be damped accurately. Modal damping formulations damp all modes specified with a modal damping accurately at all excitation frequencies. Equivalent modal damping formulations damp all modes specified only when the excitation frequency matches the natural frequency of the mode. The use of a modal viscous damping formulation to represent a modal structural damping mechanism will overestimate the modal damping when the forcing frequency ω is higher than ω_{ni} and underestimate the modal damping when the forcing frequency ω is lower than ω_{ni}. Conversely, the use of a modal structural damping formulation to represent a modal viscous damping mechanism will underestimate the modal damping when the forcing frequency ω is higher than ω_{ni} and overestimate the modal damping when the forcing frequency ω is lower than ω_{ni}. Global proportional damping will damp only one or two structural modes accurately. Global mass proportional viscous damping tends to underestimate damping in modes of higher than ω_{ni} frequencies and overestimate damping in modes of lower than ω_{ni} frequencies. Global structural damping will damp all modes accurately at all excitation frequencies if all the deformable elements exhibit structural damping of the same value. Global stiffness proportional viscous damping tends to overestimate damping in modes of higher than ω_{ni} frequencies and underestimate damping in modes of lower than ω_{ni} frequencies. Global Rayleigh damping can be made, by specifying $\zeta_r = \zeta_s$, to damp all modes between modal frequencies ω_r and ω_s to have damping less than at modes ω_r and ω_s.

1.1.3   GL, ML Base Isolation Systems

Base isolation systems are used **to minimize the effect of seismic support motion on structures and nonstructural elements** and **to reduce machine induced loadings on foundations**.

1.1.3.1   Controlling Displacement Response From Harmonic Load Excitations

We reiterate that the dynamic amplification factor $D(\omega)$ for applied harmonic loading $p_0\sin\omega t$ on a SDOF system is

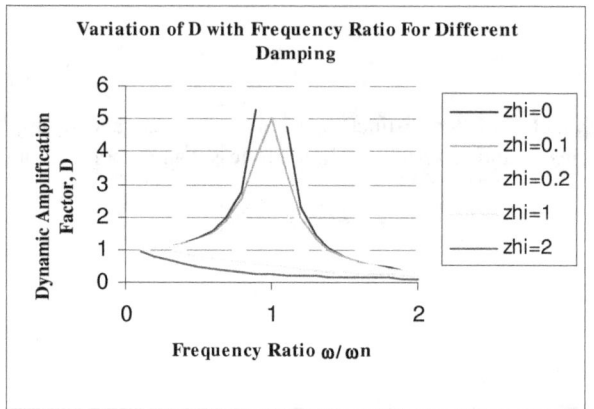

$$D_i(\omega) = \frac{1}{\sqrt{\left(1 - \omega^2/\omega_{ni}^2\right)^2 + \left(2\zeta_i\omega/\omega_{ni}\right)^2}};$$

$$D_{i\,max} = \frac{1}{2\zeta\sqrt{\left(1 - \zeta^2\right)}}$$

The corresponding maximum displacement response of the SDOF system is $u_{max} = D_{imax}(p_0/k)$. D_{imax} occurs when $\omega/\omega_n = (1-2\zeta^2)^{1/2}$. Clearly, if ω_n is flexible (i.e. low) enough such that $D(\omega)$ is less than unity (the exact location varies with the amount of damping, being at $\omega/\omega_n = 1.4$ for $\zeta = 0.0$), the steady-state dynamic response will be less than the static response and the externally applied harmonic dynamic load excitation is well controlled for the displacement response.

1.1.3.2   Controlling Acceleration Response From Harmonic Load Excitations

Now, it may also be desirable to limit the maximum acceleration amplitude. The acceleration response is simply the second derivative (with respect to time) of the displacement response. Hence the maximum acceleration response will be $a_{max} = \omega^2 D_{imax}(p_0/k) = (\omega/\omega_n)^2 D_{imax}(p_0/m)$.

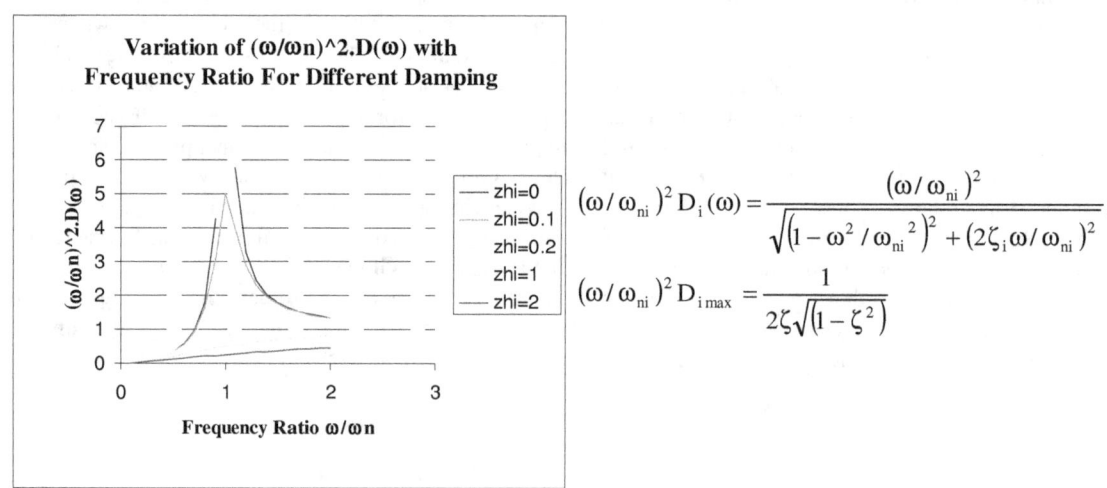

$$(\omega/\omega_{ni})^2 D_i(\omega) = \frac{(\omega/\omega_{ni})^2}{\sqrt{\left(1 - \omega^2/\omega_{ni}^2\right)^2 + \left(2\zeta_i\omega/\omega_{ni}\right)^2}};$$

$$(\omega/\omega_{ni})^2 D_{i\,max} = \frac{1}{2\zeta\sqrt{\left(1 - \zeta^2\right)}}$$

Note that the behaviour of $(\omega/\omega_n)^2 D_i(\omega)$ for small and large ω/ω_n is opposite to that of $D_i(\omega)$. The maximum is the same but the location is different. The ratio p_0/m is the acceleration of the mass if the mass is unrestrained. Hence $(\omega/\omega_n)^2 D_i(\omega)$ approaches unity for large $(\omega/\omega_n)^2$. The term $(\omega/\omega_n)^2 D_i(\omega)$ is then the multiplier to p_0/k accounting for the time varying nature of the loading and the stiffness and damping of the system. Clearly, ω/ω_n needs to be such that $(\omega/\omega_n)^2 D_i(\omega) p_0/m$ is within the acceptable acceleration limit.

1.1.3.3 Base Isolation - Controlling Displacement Response From Harmonic Base Enforced Motion

For a SDOF system subjected to base harmonic excitations $u_0 \sin\omega t$, in **absolute** terms the response would be

$$u(t) = \frac{u_0 \sqrt{1 + (2\zeta\omega/\omega_n)^2}}{\sqrt{(1 - \omega^2/\omega_n^2)^2 + (2\zeta\omega/\omega_n)^2}} \sin(\omega t + \beta - \theta) \qquad \theta = \tan^{-1}\frac{2\zeta\omega/\omega_n}{(1 - \omega^2/\omega_n^2)} \qquad \beta = \tan^{-1}(2\zeta\omega/\omega_n)$$

Defining an expression for the displacement transmissibility as the absolute displacement response amplitude divided by the amplitude of the enforcing harmonic displacement u_0,

$$T_r = \frac{\sqrt{1 + (2\zeta\omega/\omega_n)^2}}{\sqrt{(1 - \omega^2/\omega_n^2)^2 + (2\zeta\omega/\omega_n)^2}}$$

This is the **(absolute) displacement transmissibility** expression.

When we design for the displacement response, it is usually the relative response that we are concerned with. Now the steady state solution for the **relative** (to base) displacement is

$$u_r(t) = \frac{u_0 \omega^2/\omega_n^2}{\sqrt{(1 - \omega^2/\omega_n^2)^2 + (2\zeta\omega/\omega_n)^2}} \sin(\omega t - \theta) \qquad \theta = \tan^{-1}\frac{2\zeta\omega/\omega_n}{(1 - \omega^2/\omega_n^2)}$$

This expression is numerically equivalent to $(\omega/\omega_n)^2 D_i(\omega)$ and is shown below.

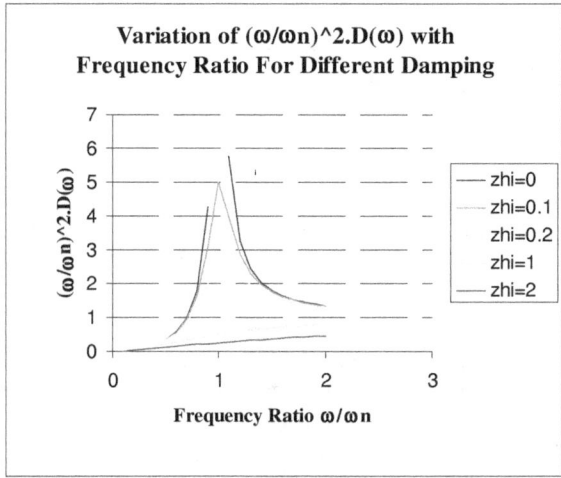

Clearly for design, ω/ω_n needs to be such that $(\omega/\omega_n)^2 D_i(\omega) u_0$ is within the acceptable relative displacement limit. Note that for large ω/ω_n values the expression approaches unity and hence the relative displacement response approaches the displacement of the base motion and needs to be designed for.

1.1.3.4   Base Isolation - Controlling Acceleration Response From Harmonic Base Enforced Motion

The **acceleration transmissibility** (absolute acceleration response amplitude divided by the amplitude of the enforcing harmonic acceleration) is exactly similar to the (absolute) displacement transmissibility.

$$T_r = \frac{\sqrt{1 + (2\zeta\omega/\omega_n)^2}}{\sqrt{(1 - \omega^2/\omega_n^2)^2 + (2\zeta\omega/\omega_n)^2}}$$

This formula is effective to determine the natural frequency (hence stiffness) and damping of an isolation system knowing the input displacement or acceleration and the maximum acceptable dynamic acceleration that the structural component can be subjected to.

Remember that only when damping is small that the maximum transmissibility can be approximated to occur at resonance. When damping is high, it is imperative that the transmissibility expression be differentiated or plotted (also use cannot be made of $\omega = \omega_n(1 - 2\zeta^2)^{1/2}$ as this is based on differentiating and maximizing the (magnitude of the) dynamic amplification factor, $D(\omega)$).

A plot of T_r versus ω/ω_n is somewhat similar to that of (magnitude of the) dynamic amplification D versus ω/ω_n, except that all the curves of different ζ pass through the same point of $T_r = 1.0$ when $\omega/\omega_n = \sqrt{2}$. Noting the curves after this point it is observed that damping tends to reduce the effectiveness of vibration isolation for frequency ratios greater than $\sqrt{2}$.

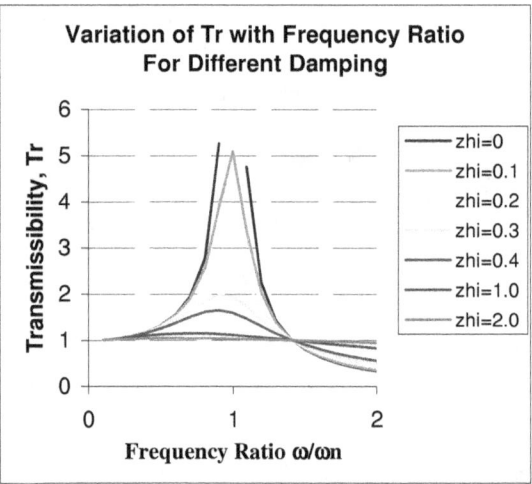

On investigating the transmissibility expression, we can design the supporting system for vibration isolation. The transmissibility is unity when the supporting system is infinitely stiff with respect to the loading frequency. The transmissibility is also unity when the frequency ratio is $\sqrt{2}$. If the supporting system is designed such that the frequency ratio is less than $\sqrt{2}$ but greater than 0, the transmissibility is greater than one, which means that the supporting system actually makes matters worse as far as vibration isolation is concerned. When the frequency ratio is greater than $\sqrt{2}$, it is seen that the transmissibility is smaller than unity and hence, the supporting system functions as a vibration isolator. Damping is seen to be advantages only in the region when the frequency ratio is less than $\sqrt{2}$ (where a spring mounting supporting system makes matters worse) and not when the frequency ratio is greater than $\sqrt{2}$ (where a spring mounting supporting system acts as a effective vibration isolator). This is not so important actually as the undesirable effects of high values of damping at frequency ratios greater than $\sqrt{2}$ is not so great especially at even higher frequency ratios, i.e. achieved by making the supporting system even more flexible.

Also, in the unfortunate circumstance that the resonance region of frequency ratio less than $\sqrt{2}$ is somehow attained; high levels of damping are extremely effective. For good isolation, it is often recommended to design the supporting system for a frequency ratio of at least 3.

An example of isolation systems in practice occurs in the case of the EUROSTAR train.

EUROSTAR BOGIE MODEL FOR DYNAMIC ANALYSIS

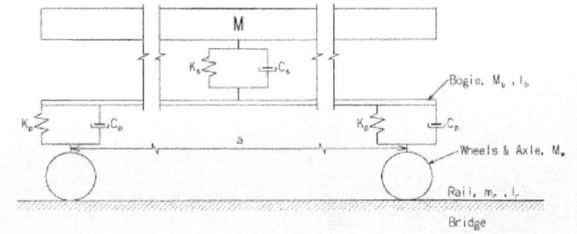

	Bogie 1	**Typical Bogie**	**Bogie 9**
M tonne	40.021	27.000	32.102
Ks kN/m	870	580	1250
Cs kNs/m	20	2	20
Mb tonne	3.025	3.250	3.200
Ib tm²	2.800	2.800	2.800
Kp kN/m	2200	2075	1560
Cp kNs/m	12	12	12
a m	3.0	3.0	3.0
Mw tonne	2.128	2.095	2.113

The primary (bogie) suspension system has a natural frequency of about 5.7Hz, hence filtering frequencies higher than $\sqrt{2} \times 5.7 = 8.1$Hz. The secondary (deck) suspension system has a natural frequency of about 0.7Hz, hence filtering frequencies higher than $\sqrt{2} \times 0.7 = 1.0$Hz.

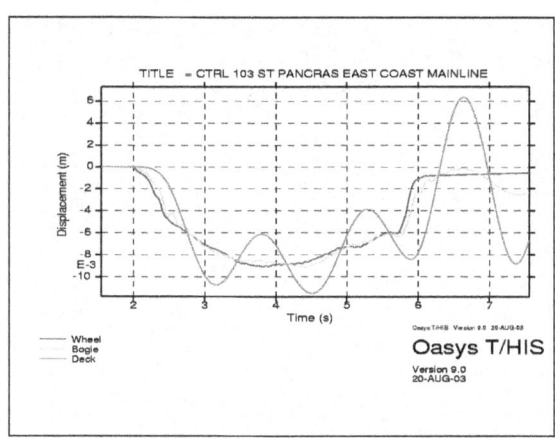

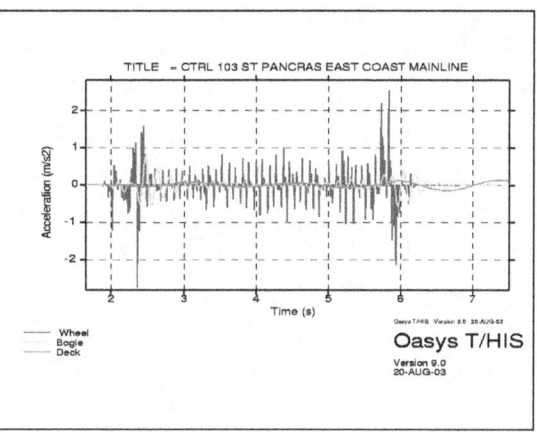

Below is the modulus FFT of the acceleration response showing clearly the dominant frequency components of the response at the wheels, bogie and train deck.

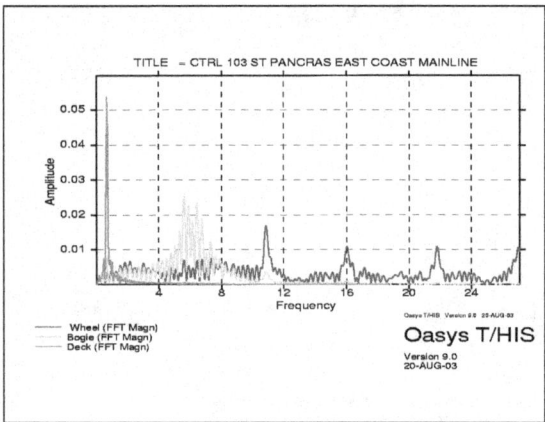

Another reason to employ isolation is to isolate **rail-induced vibrations** through the ground onto submerged structures or adjacent structures. The frequency content ranges from 1Hz to 150-200Hz. Structural vibrations occur due the 8Hz to 20Hz frequencies. Railway excitations have a pronounced frequency band between 40Hz and 80Hz, causing significant acoustic phenomena (structure borne sound). Note that noise can be detected above a frequency of about 16Hz. Vibration isolation devices typically ranging from 8Hz to 12Hz can be used to isolate frequencies above these.

Road traffic also induces vibration with a frequency range from 1 to 200-400Hz but does not, in contrast with rail-induced vibrations, exhibit pronounced frequency components.

1.1.3.5   Base Isolation – Controlling Force Transmitted Into Foundation From Harmonic Load Excitations

The **force transmissibility** (force response amplitude divided by the amplitude of the enforcing harmonic force of $p_0\sin\omega t$, i.e. p_0) is also exactly similar to the (absolute) displacement transmissibility.

$$T_r = \frac{\sqrt{1 + (2\zeta\omega/\omega_n)^2}}{\sqrt{(1 - \omega^2/\omega_n^2)^2 + (2\zeta\omega/\omega_n)^2}}$$

The transmissibility expression can be used for the force induced by a rotating machine into the supporting structure. Design supporting system such that $\omega/\omega_n = \sqrt{2}$ for no dynamic effects. Note that the transmissibility expression is only valid for the force transmitted if the foundation is rigid. If the foundation is not rigid, then the supporting system must be even more flexible in order to provide the same level of protection as when the foundation is rigid.

1.1.4   GL, ML Tuned Mass Damper (TMD) Systems

The TMD is a device consisting of a mass, spring and damper to reduce mechanical vibrations. The frequency and damping of the TMD is tuned so that when the structure resonates at a particular frequency, the TMD vibrates out-of-phase transferring the energy from the primary system (i.e. the structure) to the secondary system (i.e. the TMD) and dissipating it in the damper. Maximum flow of energy from the primary system to the TMD occurs when the displacement of the TMD lags that of the primary mass by 90°. The acceleration of the TMD is then in phase with velocity of the primary system.

The TMD concept was introduced by Frahm in 1909 to reduce the rolling motion of ships and ship hull vibrations. Significant theoretical contributions were made by Den Hartog (1940) for undamped systems subject to harmonic excitations. Extensions to the theory were then presented by Randall (1981), Warburton (1982) and Tsai & Lin (1993).

The figure below illustrates the typical configuration of a unidirectional translational tuned mass damper. The mass rests on bearings that function as rollers and allow the mass to translate laterally relative to the floor. Springs and dampers are inserted between the mass and the adjacent vertical support members which transmit the lateral "out-of-phase" force to the floor level, and then into the structural frame. Bidirectional translational dampers are configured with springs/dampers in 2 orthogonal directions and provide the capability for controlling structural motion in 2 orthogonal planes.

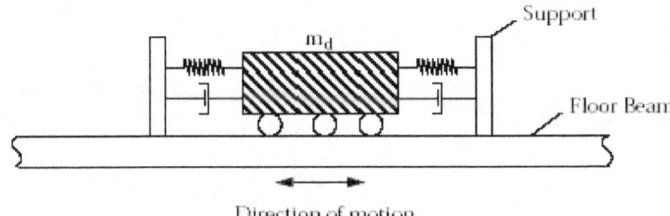

The early versions of TMD's employ complex mechanisms for the bearing and damping elements, have relatively large masses, occupy considerably space, and are quite expensive. Recent versions, such as the scheme shown below, have been designed to minimize these limitations. This scheme employs a multi-assemblage of elastomeric rubber bearings, which function as shear springs, and bitumen rubber compound (BRC) elements, which provide viscoelastic damping capability. The device is compact in size, requires unsophisticated controls, is multidirectional, and is easily assembled and modified.

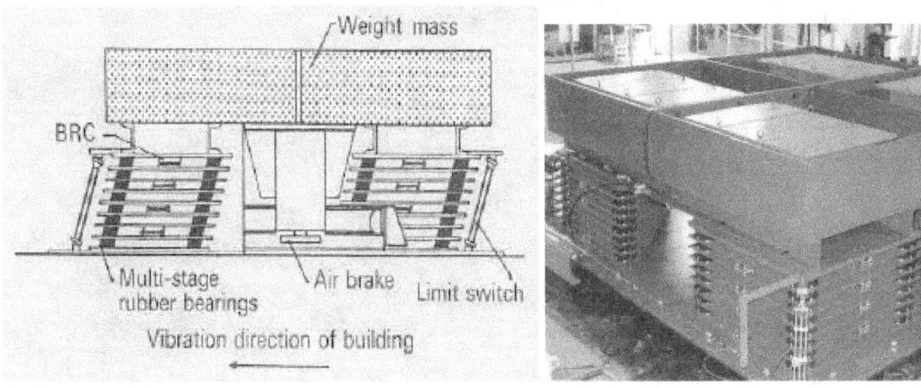

The problems associated with the bearings can be eliminated by supporting the mass with cables which allow the system to behave as a pendulum.

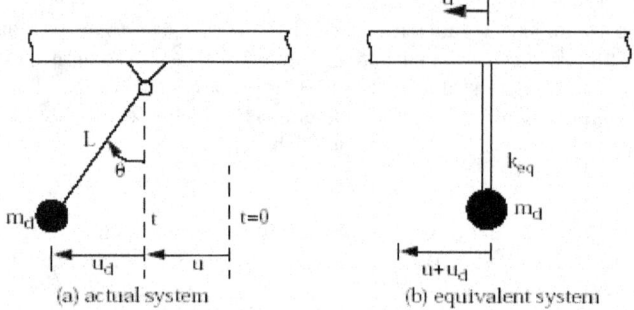

(a) actual system   (b) equivalent system

The equation of motion for the horizontal direction is

$$T \sin\theta + \frac{W_d}{g}(\ddot{u} + \ddot{u}_d) = 0$$

where T is the tension in the cable. When θ is small, the following approximations apply

$$u_d = L \sin\theta \simeq L\theta$$

$$T \simeq W_d$$

Hence

$$m_d \ddot{u}_d + \frac{W_d}{L} u_d = -m_d \ddot{u}$$

and it follows that the equivalent shear spring stiffness is

$$k_{eq} = \frac{W_d}{L}$$

The natural frequency of the pendulum is

$$\omega_d^2 = \frac{k_{eq}}{m_d} = \frac{g}{L}$$

The natural period of the pendulum is

$$T_d = 2\pi \sqrt{\frac{L}{g}}$$

The simple pendulum tuned mass damper concept has a serious limitation. Since the period depends on L, the required length for large T_d may be greater than the typical story height. This problem can be eliminated by resorting to the scheme illustrated below.

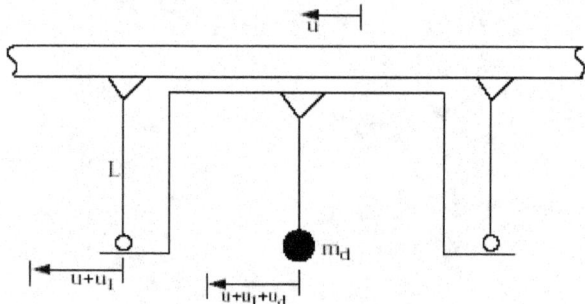

The interior rigid link magnifies the support motion for the pendulum, and results in the following equilibrium equation.

$$m_d(\ddot{u} + \ddot{u}_1 + \ddot{u}_d) + \frac{W_d}{L}u_d = 0$$

The rigid link moves in phase with the damper, and has the same displacement amplitude. Then, taking $u_1 = u_d$

$$m_d\ddot{u}_d + \frac{W_d}{2L}u_d = -\frac{m_d}{2}\ddot{u}$$

The equivalent stiffness is $W_d/2L$, and it follows that the effective length is equal to 2L. Each additional link increases the effective length by L.

The concepts of designing the passive tuned mass damper illustrated below are based on geometrically and materially linear frequency domain dynamic analyses. In deriving the equations of motion, we use absolute terms of displacement, velocity and acceleration unless we have support motion, in which case it is easier to use relative terms. Note that in absolute terms, we need the support displacement and velocity in the equation of motion whilst in relative terms we need only the support acceleration in the equation of motion.

1.1.4.1 Damped SDOF System Subject to Harmonic Force and Support Motion Excitations

The analysis of the TMD involves the prediction of the response of a two-DOF system, the first being that of the primary system and the second being the TMD. Extensive use is made of (complex arithmetic) mathematical procedures to predict the steady-state response due to harmonic excitations in the frequency domain. The employment of complex arithmetic simply allows the representation of the response and the excitation functions in complex numbers. To illustrate this elegant mathematical approach, the steady-state response of a SDOF system is first illustrated, forming the basis of further analyses involving the TMD and the primary system.

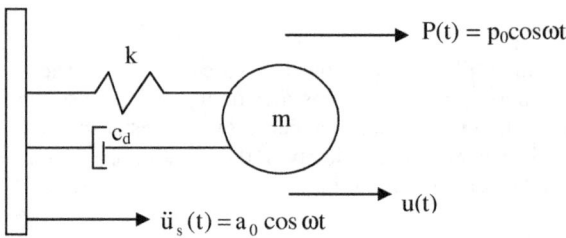

In absolute terms, the equation of motion of a SDOF system subjected to harmonic force excitation and support motion is

$$m\ddot{u}(t) + c(\dot{u}(t) - \dot{u}_s(t)) + k(u(t) - u_s(t)) = P(t)$$

In relative terms,

$$u_r(t) = u(t) - u_s(t)$$

Hence,

$$m(\ddot{u}_r(t) + \ddot{u}_s(t)) + c\dot{u}_r(t) + ku_r(t) = P(t)$$

$$m\ddot{u}_r(t) + c\dot{u}_r(t) + ku_r(t) = P(t) - m\ddot{u}_s(t)$$

For harmonic excitations,

$$P(t) = \text{Re}\,al\left[p_0 e^{i\omega t}\right] \qquad \text{and} \qquad \ddot{u}_s(t) = \text{Re}\,al\left[a_0 e^{i\omega t}\right]$$

For the inhomogenous part (steady - state solution), assume $u_r(t) = \text{Re}\,al\left[F(\omega)e^{i\omega t}\right]$,

$$-m\omega^2 F(\omega) + ic\omega F(\omega) + kF(\omega) = p_0 + ma_0$$

$$F(\omega) = \frac{p_0 + ma_0}{k - m\omega^2 + ic\omega}$$

$$F(\omega) = \frac{p_0 + ma_0}{\sqrt{(k - m\omega^2)^2 + (c\omega)^2}\,e^{i\theta}}, \theta = \tan^{-1}\frac{c\omega}{k - m\omega^2} = \tan^{-1}\frac{2\zeta\omega/\omega_n}{(1 - \omega^2/\omega_n^2)}$$

$$F(\omega) = \frac{(p_0 + ma_0)/k}{\sqrt{(1 - \omega^2/\omega_n^2)^2 + (2\zeta\omega/\omega_n)^2}}\,e^{-i\theta}, \theta = \tan^{-1}\frac{2\zeta\omega/\omega_n}{(1 - \omega^2/\omega_n^2)}$$

Thus, the steady - state solution is

$$u_r(t) = \text{Re}\,al\left[F(\omega)e^{i\omega t}\right]$$

$$u_r(t) = \frac{(p_0 + ma_0)/k}{\sqrt{(1 - \omega^2/\omega_n^2)^2 + (2\zeta\omega/\omega_n)^2}}\cos(\omega t - \theta), \theta = \tan^{-1}\frac{2\zeta\omega/\omega_n}{(1 - \omega^2/\omega_n^2)}$$

1.1.4.2   Undamped Structure, Undamped TMD System Subject to Harmonic Force Excitation

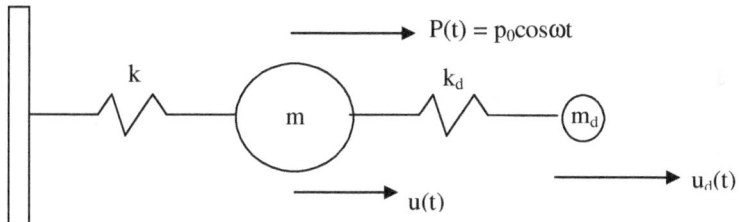

Note that
$$\omega_n{}^2 = k / m$$
$$\omega_d{}^2 = k_d / m_d$$

In absolute terms, the equations of motion of the two – DOF system subjected to harmonic force excitation are

$$m\ddot{u}(t) + ku(t) - k_d\big(u_d(t) - u(t)\big) = P(t)$$
$$m_d\ddot{u}_d(t) + k_d\big(u_d(t) - u(t)\big) = 0$$

For harmonic excitations,

$$P(t) = \mathrm{Re\,al}\big[p_0 e^{i\omega t}\big]$$

For the inhomogenous part (steady - state solution), assume $u(t) = \mathrm{Re\,al}\big[F(\omega)e^{i\omega t}\big]$ and $u_d(t) = \mathrm{Re\,al}\big[F_d(\omega)e^{i\omega t}\big]$,

$$-m\omega^2 F(\omega) + kF(\omega) - k_d F_d(\omega) + k_d F(\omega) = p_0$$
$$-m_d\omega^2 F_d(\omega) + k_d F_d(\omega) - k_d F(\omega) = 0$$

These are two equations which can be solved simultaneously for the complex frequency response functions (FRFs) $F(\omega)$ and $F_d(\omega)$ and converted into polar form to yield,

$$F(\omega) = \frac{p_0}{k} \frac{\left(f^2 - \rho^2\right)}{\left[\left(1 - \rho^2\right)\left(f^2 - \rho^2\right) - \mu\rho^2 f^2\right]} e^{i\theta_1}$$

$$F_d(\omega) = \frac{p_0}{k} \frac{\rho^2}{\left[\left(1 - \rho^2\right)\left(f^2 - \rho^2\right) - \mu\rho^2 f^2\right]} e^{i\theta_2}$$

Note that

$$f = \text{natural frequency ratio} = \frac{\omega_d}{\omega_n} = \frac{\sqrt{k_d / m_d}}{\sqrt{k / m}}$$

$$\rho = \text{forced frequency ratio} = \frac{\omega}{\omega_n} = \frac{\omega}{\sqrt{k / m}}$$

$$\mu = \text{mass ratio} = \frac{m_d}{m}$$

For completion, the steady - state solutions are

$$u(t) = \mathrm{Re\,al}\big[F(\omega)e^{i\omega t}\big]$$
$$u_d(t) = \mathrm{Re\,al}\big[F_d(\omega)e^{i\omega t}\big]$$

The plot of the magnitude of the FRF $F(\omega)$ (of the primary system) divided by the static displacement p/k, i.e. the (magnitude of the) dynamic amplification factor $D(\omega)$ is presented.

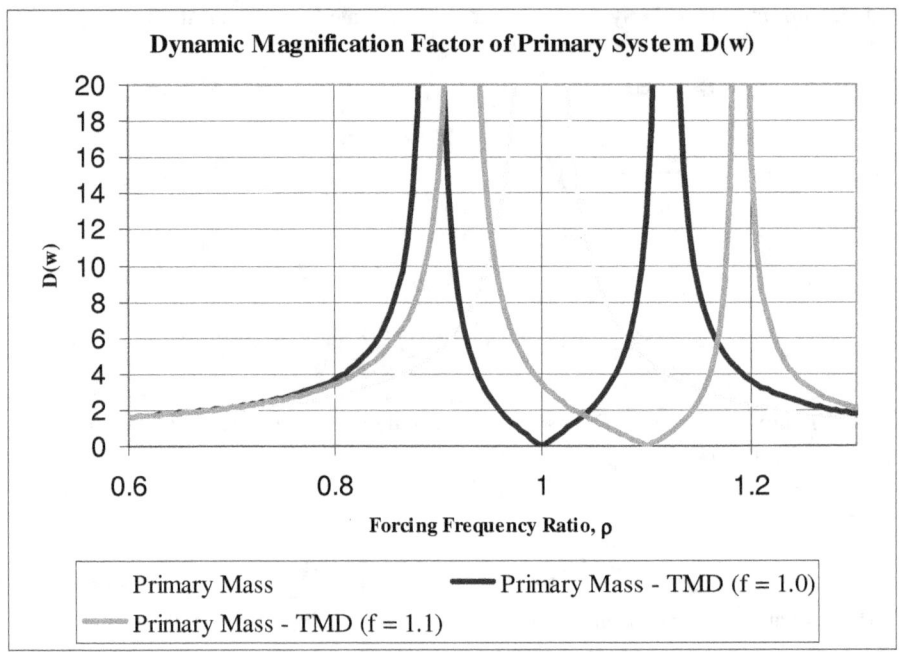

The optimum conditions are ascertained by changing the **mass ratio** μ and the **natural frequency ratio f**. Changing the frequency ratio, f has the effect of changing the position where $D(\omega)$ becomes zero, i.e. $D(\omega)$ becomes zero at when $\rho = f$. Changing the mass ratio has the effect of changing the distance between the peaks of $D(\omega)$ with higher values resulting in a greater spread.

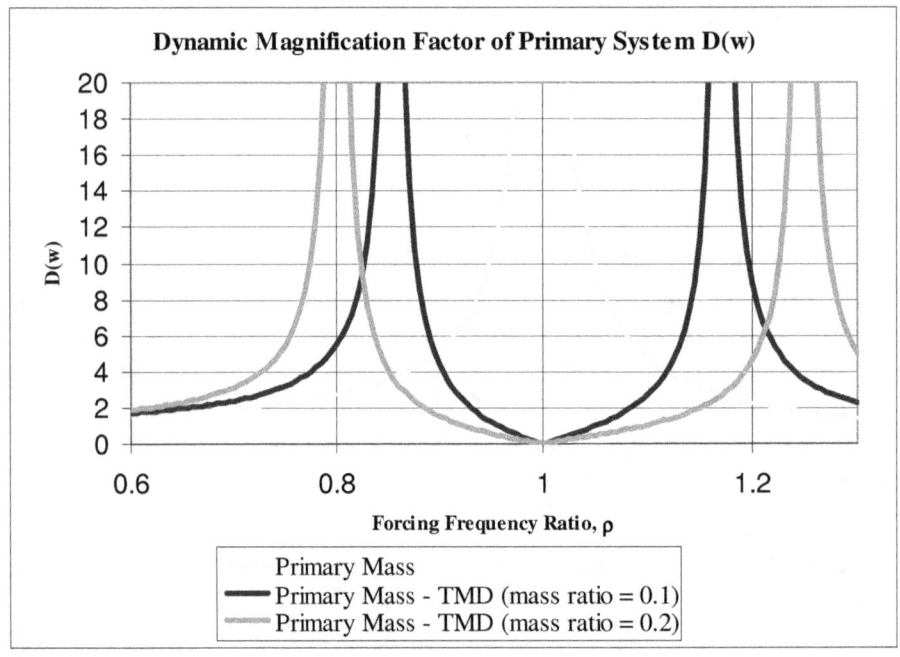

1.1.4.3 Undamped Structure, Damped TMD System Subject to Harmonic Force Excitation

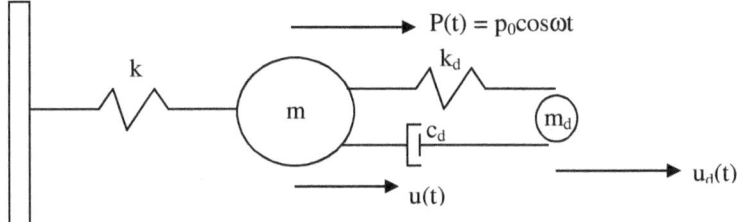

Note that
$$\omega_n^2 = k / m$$
$$\omega_d^2 = k_d / m_d$$
$$c = 2\zeta\omega_n m \text{ (here undefined)}$$
$$c_d = 2\zeta_d\omega_d m_d$$

In absolute terms, the equations of motion of the two – DOF system subjected to harmonic force excitation are
$$m\ddot{u}(t) + ku(t) - c_d(\dot{u}_d(t) - \dot{u}(t)) - k_d(u_d(t) - u(t)) = P(t)$$
$$m_d\ddot{u}_d(t) + c_d(\dot{u}_d(t) - \dot{u}(t)) + k_d(u_d(t) - u(t)) = 0$$

For harmonic excitations,
$$P(t) = \text{Re al}\left[p_0 e^{i\omega t}\right]$$

For the inhomogenous part (steady - state solution), assume $u(t) = \text{Re al}\left[F(\omega)e^{i\omega t}\right]$ and $u_d(t) = \text{Re al}\left[F_d(\omega)e^{i\omega t}\right]$,
$$-m\omega^2 F(\omega) + kF(\omega) - ic_d\omega F_d(\omega) + ic_d\omega F(\omega) - k_d F_d(\omega) + k_d F(\omega) = p_0$$
$$-m_d\omega^2 F_d(\omega) + ic_d\omega F_d(\omega) - ic_d\omega F(\omega) + k_d F_d(\omega) - k_d F(\omega) = 0$$

These are two equations which can be solved simultaneously for the complex frequency response functions (FRFs) $F(\omega)$ and $F_d(\omega)$ and converted into polar form to yield,

$$F(\omega) = \frac{p_0}{k}\frac{\sqrt{\left(f^2 - \rho^2\right)^2 + \left(2\zeta_d\rho f\right)^2}}{\sqrt{\left[\left(1 - \rho^2\right)\left(f^2 - \rho^2\right) - \mu\rho^2 f^2\right]^2 + \left[2\zeta_d\rho f\left[1 - \rho^2(1 + \mu)\right]\right]^2}}e^{i\theta_1}$$

$$F_d(\omega) = \frac{p_0}{k}\frac{\rho^2}{\sqrt{\left[\left(1 - \rho^2\right)\left(f^2 - \rho^2\right) - \mu\rho^2 f^2\right]^2 + \left[2\zeta_d\rho f\left[1 - \rho^2(1 + \mu)\right]\right]^2}}e^{i\theta_2}$$

Note that

$$f = \text{natural frequency ratio} = \frac{\omega_d}{\omega_n} = \frac{\sqrt{k_d / m_d}}{\sqrt{k / m}}$$

$$\rho = \text{forced frequency ratio} = \frac{\omega}{\omega_n} = \frac{\omega}{\sqrt{k / m}}$$

$$\mu = \text{mass ratio} = \frac{m_d}{m}$$

For completion, the steady - state solutions are
$$u(t) = \text{Re al}\left[F(\omega)e^{i\omega t}\right]$$
$$u_d(t) = \text{Re al}\left[F_d(\omega)e^{i\omega t}\right]$$

The plot of the magnitude of the FRF F(ω) (of the primary system) divided by the static displacement p/k, i.e. the (magnitude of the) dynamic amplification factor D(ω) is presented.

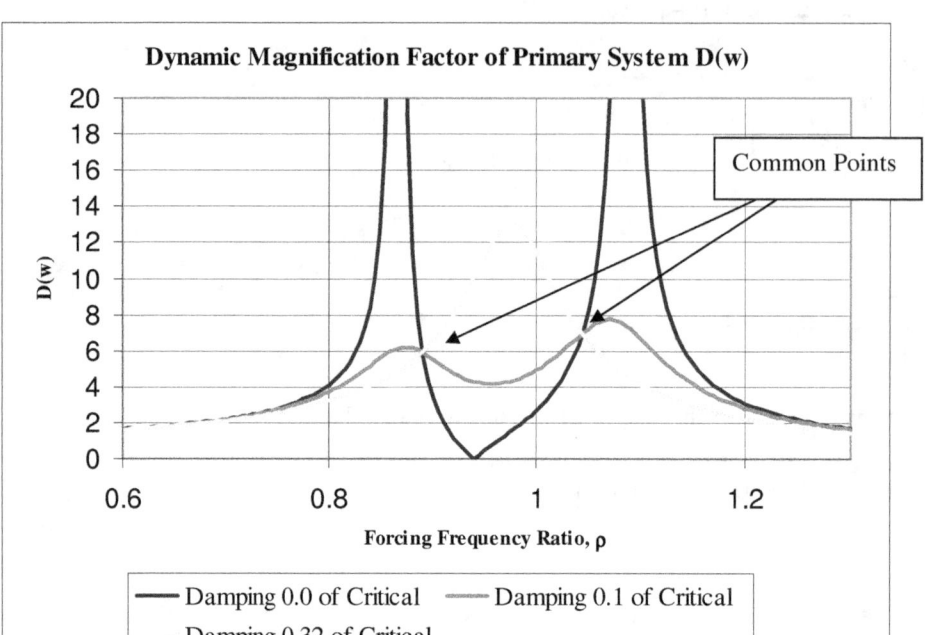

The optimum conditions are ascertained by changing the **mass ratio μ**, the **natural frequency ratio f** and the level of **damping** ζ_d. It is noticed that for whatever combination, there will always be two common points (which are independent of ζ_d) on the plot of D(ω).

The greater the mass ratio μ, the lower will be the response. Also, a larger mass ratio would render the TMD less sensitive to tuning, a characteristic which is much valuable in practice as the estimate of the modal frequencies change with insertion of cladding, non-structural components etc. The mass ratio is usually between 0.01 and 0.1.

By changing the ratio of the natural frequency ratio f, we will change the relative height of the two common points. The optimum condition will be when the two common points are at the same height. It can be shown from algebra that this condition is achieved when f is tuned such that

$$f_{opt} = \frac{1}{1+\mu}$$

which gives the amplitude at the common points as

$$D(\omega) = \sqrt{1 + \frac{2}{\mu}}$$

By changing the damping ζ_d we will change the point at which the curves attain their maxima, i.e. whether at the above common points or not. We know that the amplitude at resonance is limited only by the damping force. If $\zeta_d = 0$, there is no energy dissipation and hence two (since this is a two-DOF system) resonant peaks of infinite amplitude is attained on D(ω). When ζ_d is infinite, the two masses are practically locked to each other, their relative displacement is thus zero, hence causing no work to be done by the damper, and thus a single resonant peak of infinite amplitude is attained on D(ω). There is thus an optimum value of ζ_d for which there will be maximum work done by the damper, and this would be the point at which the response of the primary system is minimum. The optimum condition for the choice of ζ_d is achieved when the two curves pass the common points at the maxima. But this does not occur at the same time. Expressions for when the first curve passes the first common point at its maximum and the second curve passes the second common point at its maximum can be averaged to yield a useful optimum damping of

$$\zeta_{d,opt} = \sqrt{\frac{3\mu}{8(1+\mu)}}$$

Thus, in order to design a TMD to reduce vibrations from harmonic force excitations, the following approach is undertaken. Clearly, one should select the TMD location to coincide with the maximum amplitude of the mode shape that is being controlled.

(i)   Choose the practically greatest possible mass ratio, μ whilst still satisfying the maximum displacement criteria for both $F(\omega)$ and $F_d(\omega)$

(ii)   Determine the optimum natural frequency ratio, f and hence calculate the stiffness of the TMD, k_d

$$f_{opt} = \frac{1}{1+\mu}$$

$$\omega_d = f_{opt}\,\omega_n$$

$$k_d = m_d\,\omega_d^2$$

(iii)   Determine the optimum damping ratio, $\zeta_{d,opt}$ and hence calculate the optimum damping of the TMD, c_d

$$\zeta_{d,opt} = \sqrt{\frac{3\mu}{8(1+\mu)}}$$

$$c_{d,opt} = 2\zeta_{d,opt}\,\omega_d\,m_d$$

As an example, we shall investigate the design of a TMD for the London Heathrow Terminal 5 Visual Control Tower. The 85.5m high tower has a total mass of 680.9 tonnes and a fundamental global tower mode (0.91Hz) modal mass of 396 tonnes with the eigenvector **normalized at the location of the TMD**, i.e. at 78.3m. Note that there are other local (cable) modes of lower frequencies.

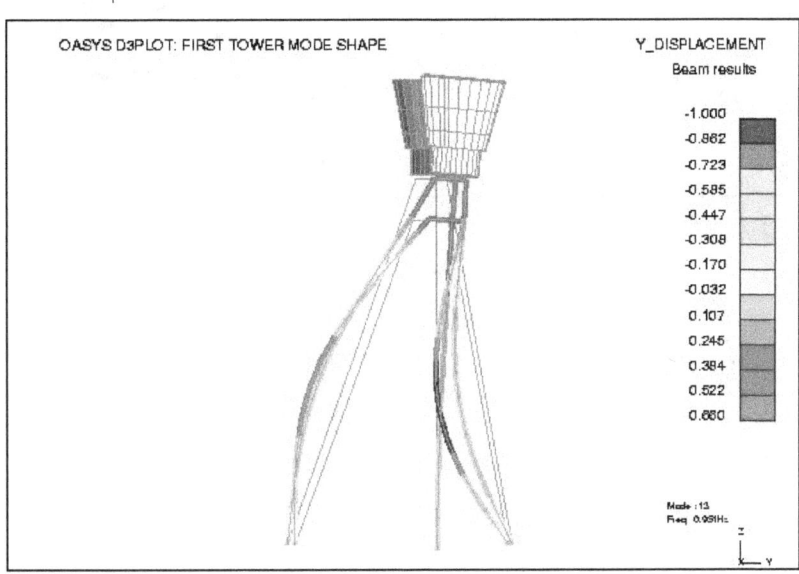

The natural frequencies of the modes of the tower are presented for completion. The corresponding modal masses are also presented with the eigenvectors MAX normalized.

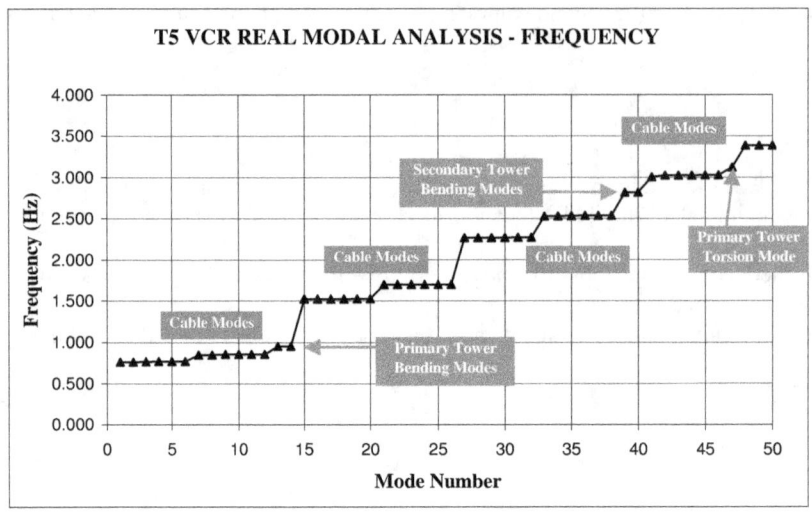

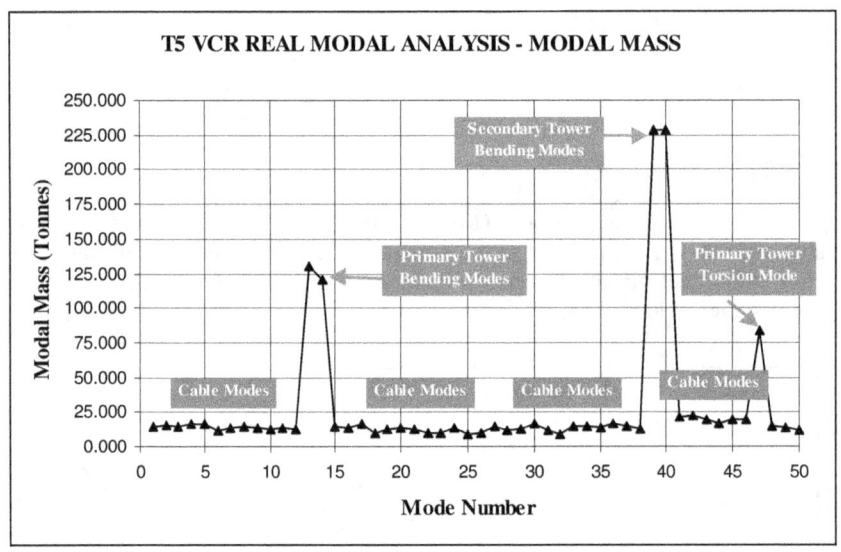

The characteristic of the modes are also expounded.

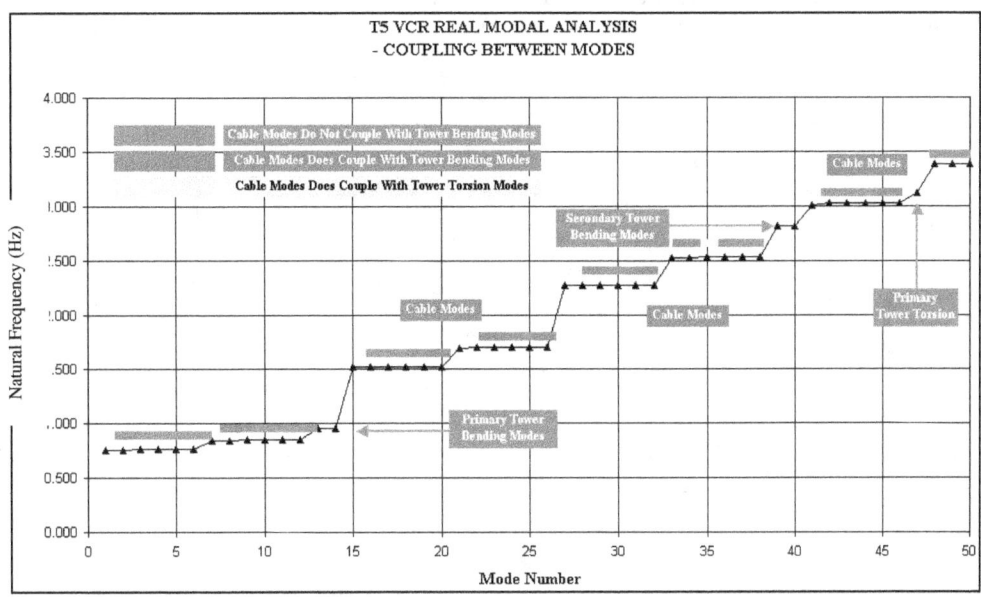

The calculation for optimum properties of a 25 tonne TMD is presented.

$$\text{Mass Ratio, } \mu = \frac{m_d}{m} = \frac{25000}{396000} = 0.0631$$

$$\text{Optimum Frequency Ratio, } f_{opt} = \frac{1}{1+\mu} = 0.9406$$

$$\text{Frequency of Secondary System, } f_d = f_{opt} f_1$$
$$= 0.9406 * 0.951$$
$$= 0.895 \text{ Hz}$$

$$\text{Optimum Stiffness of TMD, } k_d = m_d \left(2\pi f_d\right)^2$$
$$= 25000 * \left(2\pi * 0.895\right)^2$$
$$= 789745.14 \text{ N/m}$$

$$\text{Optimum Damping Ratio, } \zeta_{opt} = \sqrt{\frac{3\mu}{8(1+\mu)}}$$
$$= 0.14922$$

$$\text{Viscous Damping Coefficient, } c_d = 2 m_d \omega_d \zeta_{opt}$$
$$= 2 * 25000 * \left(2\pi f_d\right)\zeta_{opt}$$
$$= 41936.103 \text{ Ns/m}$$

The figure below shows the displacement magnification factor $D(\omega)$ at the AMD level with harmonic excitation at the AMD level. This shows that the *effective* damping of the primary tower mode from the TMD is 9.7% of critical since the reduction of maximum amplification reduces from 33.2 to 4.84.

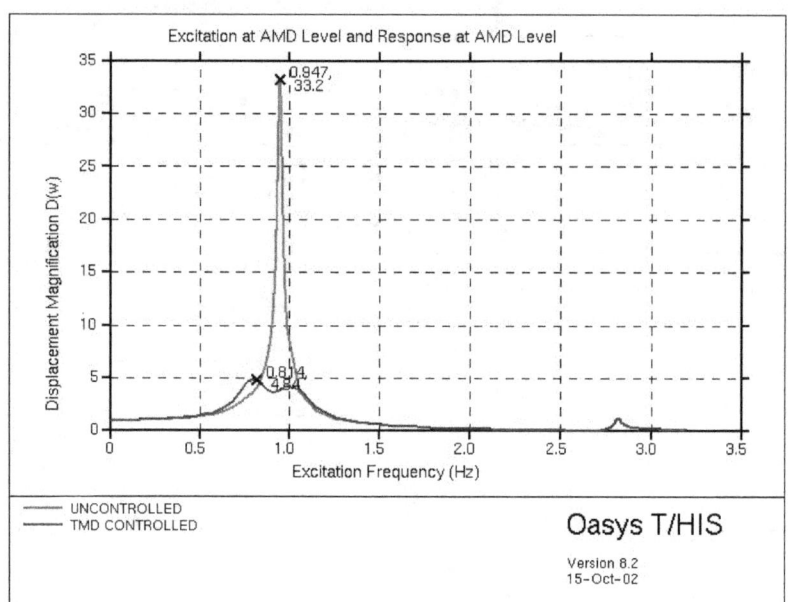

1.1.4.4 Undamped Structure, Damped TMD System Subject to Harmonic Support Motion

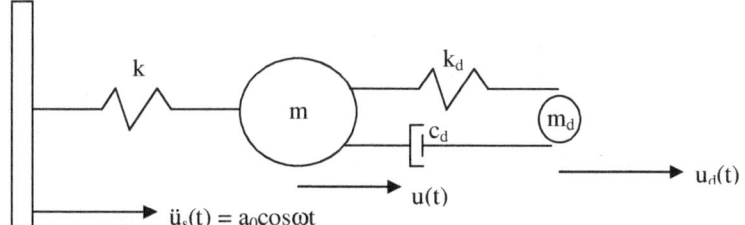

In absolute terms, the equations of motion of the two – DOF system subjected to harmonic support motion are

$$m\ddot{u}(t) + k\big(u(t) - u_s(t)\big) - c_d\big(\dot{u}_d(t) - \dot{u}(t)\big) - k_d\big(u_d(t) - u(t)\big) = 0$$

$$m_d\ddot{u}_d(t) + c_d\big(\dot{u}_d(t) - \dot{u}(t)\big) + k_d\big(u_d(t) - u(t)\big) = 0$$

In relative terms,

$$u_r(t) = u(t) - u_s(t)$$

$$u_{dr}(t) = u_d(t) - u(t) = u_d(t) - u_r(t) - u_s(t)$$

Hence,

$$m\ddot{u}_r(t) + k u_r(t) - c_d\dot{u}_{dr}(t) - k_d u_{dr}(t) = -m\ddot{u}_s(t)$$

$$m_d\ddot{u}_{dr}(t) + c_d\dot{u}_{dr}(t) + k_d u_{dr}(t) = -m_d\ddot{u}_s(t) - m_d\ddot{u}_r(t)$$

For harmonic excitations,

$$\ddot{u}_s(t) = \mathrm{Re\,al}\big[a_0 e^{i\omega t}\big]$$

For the inhomogenous part (steady - state solution), assume $u_r(t) = \mathrm{Re\,al}\big[F(\omega)e^{i\omega t}\big]$ and $u_{dr}(t) = \mathrm{Re\,al}\big[F_d(\omega)e^{i\omega t}\big]$,

$$-m\omega^2 F(\omega) + kF(\omega) - ic_d\omega F_d(\omega) - k_d F_d(\omega) = -ma_0$$

$$-m_d\omega^2 F_d(\omega) + ic_d\omega F_d(\omega) + k_d F_d(\omega) = -m_d a_0 + m_d\omega^2 F(\omega)$$

These are two equations which can be solved simultaneously for the complex frequency response functions (FRFs) $F(\omega)$ and $F_d(\omega)$ and converted into polar form to yield,

$$F(\omega) = -\frac{a_0 m}{k}\frac{\sqrt{\big[(1+\mu)f^2 - \rho^2\big]^2 + (2\zeta_d\rho f(1+\mu))^2}}{\sqrt{\big[(1-\rho^2)(f^2-\rho^2) - \mu\rho^2 f^2\big]^2 + \big[2\zeta_d\rho f\big[1-\rho^2(1+\mu)\big]\big]^2}}\, e^{i\theta_3}$$

$$F_d(\omega) = -\frac{a_0 m}{k}\frac{1}{\sqrt{\big[(1-\rho^2)(f^2-\rho^2) - \mu\rho^2 f^2\big]^2 + \big[2\zeta_d\rho f\big[1-\rho^2(1+\mu)\big]\big]^2}}\, e^{i\theta_4}$$

Note that

$$f = \text{natural frequency ratio} = \frac{\omega_d}{\omega_n} = \frac{\sqrt{k_d/m_d}}{\sqrt{k/m}}$$

$$\rho = \text{forced frequency ratio} = \frac{\omega}{\omega_n} = \frac{\omega}{\sqrt{k/m}}$$

$$\mu = \text{mass ratio} = \frac{m_d}{m}$$

For completion, the steady - state solutions are

$$u_r(t) = \mathrm{Re\,al}\big[F(\omega)e^{i\omega t}\big]$$

$$u_{dr}(t) = \mathrm{Re\,al}\big[F_d(\omega)e^{i\omega t}\big]$$

The amplification factors for harmonic ground motion are essentially similar to the amplification factors for external harmonic loading so long as the mass ratio is small. The exact expressions for the optimum conditions for harmonic ground motion are nevertheless presented here.

As before, the optimum conditions are ascertained by changing the mass ratio μ, the forcing frequency ratio f and the level of damping ζ_d. It is noticed that for whatever combination, there will always be two common points (which are independent of ζ_d) on the plot of $D(\omega)$.

The greater the mass ratio, the lower will be the response. The mass ratio is usually between 0.01 and 0.1.

By changing the ratio of the forcing frequency f, we will change the relative height of the two common points. The optimum condition will be when the two common points are at the same height. The optimum frequency ratio f is

$$f_{opt} = \frac{\sqrt{1 - \mu/2}}{1 + \mu}$$

The optimum condition for the choice of ζ_d is achieved when the two curves pass the common points at the maxima. But this does not occur at the same time. An average damping ratio that yields the optimum damping is

$$\zeta_{d,opt} = \sqrt{\frac{3\mu}{8(1 + \mu)(1 - \mu/2)}}$$

Thus, in order to design a TMD to reduce vibrations from harmonic force excitations, the following approach is undertaken. Clearly, one should select the TMD location to coincide with the maximum amplitude of the mode shape that is being controlled.

(i)   Choose the practically greatest possible mass ratio, μ whilst still satisfying the maximum displacement criteria for both $F(\omega)$ and $F_d(\omega)$

(ii)   Determine the optimum natural frequency ratio, f and hence calculate the stiffness of the TMD, k_d

$$f_{opt} = \frac{\sqrt{1 - \mu/2}}{1 + \mu}$$
$$\omega_d = f_{opt} \omega_n$$
$$k_d = m_d \omega_d^2$$

(iii)   Determine the optimum damping ratio, $\zeta_{d,opt}$ and hence calculate the optimum damping of the TMD, c_d

$$\zeta_{d,opt} = \sqrt{\frac{3\mu}{8(1 + \mu)(1 - \mu/2)}}$$
$$c_{d,opt} = 2\zeta_{d,opt} \omega_d m_d$$

1.1.4.5 Damped Structure, Damped TMD System Subject to Harmonic Force and Support Motion

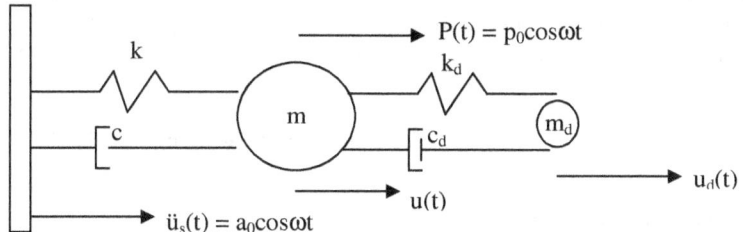

In absolute terms, the equations of motion of the two – DOF system subjected to harmonic force and support motion are

$$m\ddot{u}(t) + c\big(\dot{u}(t) - \dot{u}_s(t)\big) + k\big(u(t) - u_s(t)\big) - c_d\big(\dot{u}_d(t) - \dot{u}(t)\big) - k_d\big(u_d(t) - u(t)\big) = P(t)$$

$$m_d\ddot{u}_d(t) + c_d\big(\dot{u}_d(t) - \dot{u}(t)\big) + k_d\big(u_d(t) - u(t)\big) = 0$$

In relative terms,

$$u_r(t) = u(t) - u_s(t)$$

$$u_{dr}(t) = u_d(t) - u(t) = u_d(t) - u_r(t) - u_s(t)$$

Hence,

$$m\ddot{u}_r(t) + c\dot{u}_r(t) + ku_r(t) - c_d\dot{u}_{dr}(t) - k_d u_{dr}(t) = P(t) - m\ddot{u}_s(t)$$

$$m_d\ddot{u}_{dr}(t) + c_d\dot{u}_{dr}(t) + k_d u_{dr}(t) = -m_d\ddot{u}_s(t) - m_d\ddot{u}_r(t)$$

For harmonic excitations,

$$P(t) = \mathrm{Re\,al}\big[p_0 e^{i\omega t}\big] \qquad \text{and} \qquad \ddot{u}_s(t) = \mathrm{Re\,al}\big[a_0 e^{i\omega t}\big]$$

For the inhomogenous part (steady - state solution), assume $u_r(t) = \mathrm{Re\,al}\big[F(\omega)e^{i\omega t}\big]$ and $u_{dr}(t) = \mathrm{Re\,al}\big[F_d(\omega)e^{i\omega t}\big]$,

$$-m\omega^2 F(\omega) + ic\omega F(\omega) + kF(\omega) - ic_d\omega F_d(\omega) - k_d F_d(\omega) = p_0 - ma_0$$

$$-m_d\omega^2 F_d(\omega) + ic_d\omega F_d(\omega) + k_d F_d(\omega) = -m_d a_0 + m_d\omega^2 F(\omega)$$

These are two equations which can be solved simultaneously for the complex frequency response functions (FRFs) $F(\omega)$ and $F_d(\omega)$ and converted into polar form to yield,

$$F(\omega) = \frac{p_0}{k}\frac{\sqrt{\big(f^2 - \rho^2\big)^2 + \big(2\zeta_d\rho f\big)^2}}{\Big[-f^2\rho^2\mu + \big(1-\rho^2\big)\big(f^2-\rho^2\big) - 4\zeta\zeta_d f\rho^2\Big]^2 + 4\Big[\zeta\rho\big(f^2-\rho^2\big) + \zeta_d f\rho\big[1-\rho^2(1+\mu)\big]\Big]^2}e^{i\theta_1}$$

$$-\frac{a_0 m}{k}\frac{\sqrt{\big[(1+\mu)f^2 - \rho^2\big]^2 + \big(2\zeta_d\rho f(1+\mu)\big)^2}}{\Big[-f^2\rho^2\mu + \big(1-\rho^2\big)\big(f^2-\rho^2\big) - 4\zeta\zeta_d f\rho^2\Big]^2 + 4\Big[\zeta\rho\big(f^2-\rho^2\big) + \zeta_d f\rho\big[1-\rho^2(1+\mu)\big]\Big]^2}e^{i\theta_2}$$

$$F_d(\omega) = \frac{p_0}{k}\frac{\rho^2}{\Big[-f^2\rho^2\mu + \big(1-\rho^2\big)\big(f^2-\rho^2\big) - 4\zeta\zeta_d f\rho^2\Big]^2 + 4\Big[\zeta\rho\big(f^2-\rho^2\big) + \zeta_d f\rho\big[1-\rho^2(1+\mu)\big]\Big]^2}e^{i\theta_3}$$

$$-\frac{a_0 m}{k}\frac{1}{\Big[-f^2\rho^2\mu + \big(1-\rho^2\big)\big(f^2-\rho^2\big) - 4\zeta\zeta_d f\rho^2\Big]^2 + 4\Big[\zeta\rho\big(f^2-\rho^2\big) + \zeta_d f\rho\big[1-\rho^2(1+\mu)\big]\Big]^2}e^{i\theta_4}$$

For completion, the steady - state solutions are

$$u_r(t) = \mathrm{Re\,al}\big[F(\omega)e^{i\omega t}\big]$$

$$u_{dr}(t) = \mathrm{Re\,al}\big[F_d(\omega)e^{i\omega t}\big]$$

As before, the optimum conditions are ascertained by changing the mass ratio μ, the forcing frequency ratio f and the level of damping ζ_d. The greater the mass ratio, the lower will be the response. The mass ratio is usually between 0.01 and 0.1. Analytical solutions for the optimum natural frequency ratio, f and the optimum damping ratio, $\zeta_{d,opt}$ cannot be found. The two common points (which are independent of ζ_d) on the $D(\omega)$ plot no longer exist.

For force excitation, the optimum values of f and ζ_d for small values of ζ can be formulated empirically (Ioi and Ikeda, 1978) as follows.

$$f_{opt} = \frac{1}{1+\mu} - (0.241 + 1.7\mu - 2.6\mu^2)\zeta - (1.0 - 1.9\mu + \mu^2)\zeta^2$$

$$\zeta_d = \sqrt{\frac{3\mu}{8(1+\mu)}} + (0.13 + 0.12\mu + 0.4\mu^2)\zeta - (0.01 + 0.9\mu + 3\mu^2)\zeta^2$$

Tolerance or permissible error ranges attached to the above equations are found to be less than 1% for $0.03 < \mu < 0.40$ and $0.0 < \zeta < 0.15$, which are the ranges of practical interest.

1.1.4.6 TMD Analysis Summary (Warburton, 1982)

Clearly, one should select the TMD location to coincide with the maximum amplitude of the mode shape that is being controlled. The mass of the primary system will be the (POINT normalized to the TMD DOF) modal mass of the mode in which damping is intended. The greater the mass ratio μ, the lower will be the response and the smaller the sensitivity to tuning. Hence, choose the practically greatest possible mass ratio, μ whilst still satisfying the maximum displacement criteria for both $F(\omega)$ and $F_d(\omega)$. Then determine the optimum natural frequency ratio, f and hence calculate the stiffness of the TMD, $k_d = m_d\omega_d^2$. Finally, determine the optimum damping ratio, $\zeta_{d,opt}$ and hence calculate the optimum damping of the TMD, $c_d = 2\zeta_{d,opt}\omega_d m_d$. The structure and the TMD can be modeled and analyzed within a MSC.NASTRAN frequency domain analysis (SOL 108 or SOL 112) to confirm the results.

System	Excitation Type	Excitation Applied To	Optimized Parameter	Optimum Natural Frequency Ratio, $f_{opt} = \omega_d/\omega_n$	Optimum TMD Damping, $\zeta_{d,opt}$	Optimized Maximum (Relative) Response of Structure, u_r
Damped System	Deterministic Harmonic	Primary Mass and Support	Primary Mass Displacement	N/A	N/A	$\dfrac{(p_0 + ma_0)/k}{\sqrt{2\zeta(1-\zeta^2)}}$
Undamped System, Undamped TMD	Deterministic Harmonic	Primary Mass	Primary Mass Displacement	N/A	N/A	0.0
Undamped System, Damped TMD	Deterministic Harmonic	Primary Mass	Primary Mass Displacement	$\dfrac{1}{1+\mu}$	$\sqrt{\dfrac{3\mu}{8(1+\mu)}}$	$\dfrac{p_0}{k}\sqrt{1+\dfrac{2}{\mu}}$
Undamped System, Damped TMD	Deterministic Harmonic	Primary Mass	Primary Mass Acceleration	$\dfrac{1}{\sqrt{1+\mu}}$	$\sqrt{\dfrac{3\mu}{8(1+\mu/2)}}$	$\dfrac{p_0}{k}\sqrt{\dfrac{2}{\mu(1+\mu)}}$
Undamped System, Damped TMD	Deterministic Harmonic	Support	Primary Mass Displacement	$\dfrac{\sqrt{1-\mu/2}}{1+\mu}$	$\sqrt{\dfrac{3\mu}{8(1+\mu)(1-\mu/2)}}$	$\dfrac{ma_0}{k}(1+\mu)\sqrt{\dfrac{2}{\mu}}$
Undamped System, Damped TMD	Deterministic Harmonic	Support	Primary Mass Acceleration	$\dfrac{1}{1+\mu}$	$\sqrt{\dfrac{3\mu}{8(1+\mu)}}$	$\dfrac{ma_0}{k}\sqrt{1+\dfrac{2}{\mu}}$
Undamped System, Damped TMD	Random	Primary Mass		$\dfrac{\sqrt{1+\mu/2}}{1+\mu}$	$\sqrt{\dfrac{\mu(1+3\mu/4)}{4(1+\mu)(1+\mu/2)}}$	$\dfrac{p_0}{k}\sqrt{\dfrac{1+3\mu/4}{\mu(1+\mu)}}$
Undamped System, Damped TMD	Random	Support		$\dfrac{\sqrt{1-\mu/2}}{1+\mu}$	$\sqrt{\dfrac{\mu(1-\mu/4)}{4(1+\mu)(1-\mu/2)}}$	$\dfrac{ma_0}{k}(1+\mu)^{1.5}\cdot\sqrt{\dfrac{1}{\mu}-\dfrac{1}{4}}$
Damped System, Damped TMD	Deterministic Harmonic	Primary Mass	Primary Mass Displacement	$\dfrac{1}{1+\mu}$ $-(0.241+1.7\mu-2.6\mu^2)\zeta$ $-(1.0-1.9\mu+\mu^2)\zeta^2$	$\sqrt{\dfrac{3\mu}{8(1+\mu)}}$ $+(0.13+0.12\mu+0.4\mu^2)\zeta$ $-(0.01+0.9\mu+3\mu^2)\zeta^2$	
Damped System, Damped TMD	Deterministic Harmonic	Primary Mass	Primary Mass Displacement	$\dfrac{1}{\sqrt{1+\mu}}$ $+(0.096+0.88\mu-1.8\mu^2)\zeta$ $+(1.34-2.9\mu+3\mu^2)\zeta^2$	$\sqrt{\dfrac{3\mu(1+0.49\mu-0.2\mu^2)}{8(1+\mu)}}$ $+(0.13+0.72\mu+0.2\mu^2)\zeta$ $+(0.19+1.6\mu-4\mu^2)\zeta^2$	

The displacement optimization parameter is used essentially to determine the safety and integrity of the structure under external excitations. Since large accelerations of a structure under excitations produce detrimental effects on functionality of nonstructural components, base shear and occupant comfort, minimizing structural accelerations can also be a viable optimization criterion.

The effective damping often quoted for a system with a TMD is obtained simply by the following equation

$$\zeta_{eff} = 1/(2D_{opt}(\omega))$$

as this is consistent with the definition of the maximum dynamic amplification for SDOF systems without TMD which for small values of ζ is

$$D_{max}(\omega) = 1/(2\zeta)$$

For broadband seismic applications, Villaverde (1985) had suggested the following equations for optimum parameters with mass ratio based on modal mass with eigenvector POINT normalized to location of TMD.

$$f_{opt} = 1$$

$$\zeta_{d,opt} = \zeta + \mu$$

For broadband seismic applications, Fadek et. al. (1997) suggested the following equations for optimum parameters with mass ratio based on modal mass with eigenvector POINT normalized to location of TMD.

$$f_{opt} = \frac{1}{1+\mu}\left[1 - \zeta\sqrt{\frac{\mu}{1+\mu}}\right]$$

$$\zeta_{d,opt} = \frac{\zeta}{1+\mu} + \sqrt{\frac{\mu}{1+\mu}}$$

TMD are effective in seismic applications for structures with low damping, $\zeta = 0.02$. For structures with damping, $\zeta = 0.05$, TMDs are not very effective since a very large mass ratio is required. TMDs are also not effective in seismic applications of very stiff structures with periods 0.1-0.2s.

1.2   GL, ML Active Structural Motion Control – Control System Analysis

A control system provides feedback (output) to an input. This is useful to model an Active Mass Damper (AMD). The modelling of a control system involves the definition of an input-output relationship, i.e. a transfer function. This can be defined using the TF bulk data entry. Nonlinearities can be simulated via the NOLINi entries with which nonlinear transient loads are expressed as functions of displacements or velocities. Complex eigenvalue analysis can be used to determine stability when control systems include damping and unsymmetrical matrices.

BIBLIOGRAPHY

1.   CONNOR, Jerome. *Introduction to Structural Motion Control.* Prentice Hall, Massachusetts Institute of Technology, 2003.
2.   BACHMANN, Hugo. *Vibration Problems in Structures Practical Guidelines.* Birkhauser, Berlin, 1995.
3.   EWINS, D.J. *Modal Testing Theory, Practice and Application Second Edition.* Research Studies Press Ltd, England, 2000.
4.   JIMIN, He and ZHI-FANG, Fu. *Modal Analysis.* Butterworth-Heinemann, Great Britain, 2001.
5.   BRUEL & KJAER. *Vibration Analysis, Measurement and Testing.* Denmark, 2001.
6.   CLOUGH R.W. & PENZIEN J. *Dynamics of Structures 2nd Edition.* McGraw-Hill Book Co., United States, 1993.
7.   SMITH, J.W. *Vibration of Structures, Application in Civil Engineering Design.* Chapman and Hall, London, 1988.
8.   BEARDS, C. F. *Structural Vibration Analysis, Modelling, Analysis and Damping in Vibrating Structures.* Ellis Horwood, London, 1983.
9.   SELVAM, Dr. R.P & GOVINDASWAMY, Suresh. *Aeroelastic Analysis of Bridge Girder Section Using Computer Modelling.* University of Arkansas, May 2001.
10. SIMIU and SCALAN. *Wind Effects on Structures, An Introduction to Wind Engineering.*

www.ingramcontent.com/pod-product-compliance
Lightning Source LLC
Chambersburg PA
CBHW081304170526
45165CB00011B/3411